Reliability is a New Science

Paolo Rocchi

Reliability is a New Science

Gnedenko was Right

With an Appendix by Gurami Sh. Tsitsiashvili

Paolo Rocchi
IBM
Rome
Italy

and

LUISS Guido Carli University
Rome
Italy

ISBN 978-3-319-86171-5 ISBN 978-3-319-57472-1 (eBook)
DOI 10.1007/978-3-319-57472-1

Softcover reprint of the hardcover 1st edition 2017

Printed on acid-free paper

This Springer imprint is published by Springer Nature
The registered company is Springer International Publishing AG
The registered company address is: Gewerbestrasse 11, 6330 Cham, Switzerland

On fait la science avec des faits, comme on fait une maison avec des pierres: mais une accumulation de faits n'est pas plus une science qu'un tas de pierres n'est une maison.

Henri Poincaré

Contents

About the Author

Paolo Rocchi has been recognized as an Emeritus Docent at IBM for his achievements, and serves as an Adjunct Professor at Luiss University of Rome. He is still active in various areas of research, including information theory, probability theory, computer science education, and software engineering. Rocchi is a solitary 'pathfinder' who is no stranger to groundbreaking proposals. He has authored over one hundred and thirty works including a dozen volumes.

Acronyms

AAP	Advanced Aging Period
ABAO	As-Bad-As-Old
AGAN	As-Good-As-New
ALT	Accelerated Life Testing
CF	Continuous Factors
CSE	Continuous Side Effects
DfR	Design for Reliability
DRM	Degradation Reliability Methods
IDL	International Database on Longevity
MOF	Multi-Organ Failure
MTBF	Mean Time Between Failure
MTTF	Mean Time To Failure
NAP	Normal Aging Period
NTA	Network Theory of Aging
PHM	Proportional Hazard Model
POF	Physics of Failure
PM	Preventive Maintenance
RF	Random Factors
TS	Thermodynamic System

Introduction

Research in the reliability theory was boosted after the Second World War when the industrial production registered a significant rise in both the capitalist and communist economies.

Nowadays, companies develop new, higher technology products in record time while improving quality and productivity, and mathematicians are called to sustain such commercial strategies. They strive to set up rigorous methods of calculus for the purpose of improving the dependability of products and services. A large amount of results have been gained in the reliability territory yet the book "*Математические Методы в Теории Надёжности* (*Mathematical Methods in Reliability Theory*)" written by Boris V. Gnedenko, Alexander D. Soloviev, and Yuri K. Bélyaev in 1966 still towers as a pillar of this sector's configuration and identity. The authors conceived *reliability as a new science* and openly spoke their mind in this passage:

> It is undeniable that the reliability theory is a complex science, directly related primarily to the competence of the engineer, the physicist, the economist and chemist (…) the study of the peculiarities of biological systems from the viewpoint of the reliability theory will discover new principles.

The Russian masters were not confined to words, and they also laid the first stones to build up this scientific domain and achieved significant outcomes. Notably they took the first steps to create reliability as a new discipline.

Chapter 1
Background and Motivation

Since time immemorial ensuring the dependability of machines and buildings was felt to be a normal duty by engineers. Manufacturers systematically made sure that mechanisms created by them would operate when expected. In the medical field, the basic service offered by doctors was to ensure good health to the patient. In literature and in professional practices one can see a long term concern regarding good-functioning systems—animate as well inanimate—but the concept of 'reliability' did not have scientific status for centuries and emerged with a precise meaning only in the twenties of the past century when it was used in connection with various industrial issues. The development of products was then on a parallel path with quality while the *reliability theory* took the first steps.

During the Second World War there was great interest in weapons and other equipment that would be able to work without interruption in prolonged fighting and under extreme conditions. Theorists supported this strategic goal with extensive inquiries and began to arrange a mathematical corpus of knowledge about the reliability of systems where probability calculus and statistical methods played key roles.

The high productivity trends caused by the Second World War continued after the end of the conflict, the so-called *postwar boom* was a period of economic progress in the capitalist and communist countries alike. Significant contributions to the reliability theory came from the US and URSS. In some sense the two countries, competing in the Cold War, became two cooperative poles of research in this domain. Governments and private industries had to recognize the importance of understanding and strengthening the theoretical methods. Researches were given much impetus, the number of papers increased as steadily as Ushakov describes the period between the late 50s and the early 60s as the time of the reliability 'Renaissance'.

In the subsequent decades, manufacturers moved toward various directions to win challenging markets. They accelerated product development, and searched for new technologies and materials. They devised sophisticated products of higher quality. All this stimulated a variety of streams of studies and the researches in the

P. Rocchi, *Reliability is a New Science*, DOI 10.1007/978-3-319-57472-1_1

dependability area have gone through significant transformations during the course of its relatively short history. Ushakov [1] identifies the following main veins of research in this dynamic context:

1. Quality Control of Mass Production,
2. Reliability Engineering,
3. Reliability Testing,
4. Pure Theoretical Studies.

Appendix A offers a bibliographical selection of books, essays and articles that deal with the history of the reliability researches and reveal some motivations emerging in the scientific community. The various accounts prove without a doubt how experts have aimed to provide effective answers to the practical issues arising from companies and institutions while general statements have not been so interesting for them. Pure theoretical studies 4 make up a minor portion. Mathematicians have mostly been attracted by applied problems and sometimes the reliability theory looks like a 'heap' of works dealing with a broad assortment of topics that sometimes do not have clear connections in terms of logic.

Henri Poincaré writes: "*Science is built up with facts, as a house is made of stones: but a collection of facts is no more a science than a heap of stones is a house*" [2]. A discipline consists of cultural 'stones' and Poincaré emphasizes the difference between a 'heap of stones' and a 'house' whose stones are regularly connected. The great amount of data-driven inquiries about system dependability can be compared to an aggregate of things placed randomly one on top of the other because of the missing comprehensive frame. The metaphor of Poincaré could be applied to the present domain which has not yet matured into a science even if progress in this direction would be desirable from many viewpoints. In fact, the *reliability science* would facilitate the engineering methods of work; it would accelerate the discovery of new solutions; it would make easier technology advances and provide many other advantages with respect to the present assortmente of applications. The reliability territory crosses several areas from mechanical engineering to biology, from electronics to medicine; thus the benefits promised by the reliability science would increase in multiple ways. When the present sector will become a discipline on par with the physics and chemistry, scholars will be able to treat machines and people, technology and biology within a unifying frame of knowledge.

It is not difficult to forecast that the evolution of the present domain toward a mature status will require a long process. Reasonably one could say that the advance of this field will have to go through three main stages: the present stage which scholars call the *reliability theory*, a second stage to which we give the name of *principle-based reliability theory* and eventually the stage of *reliability science*.

Hence the principle-based reliability theory is just waiting to be set up.

1.1 What Is a Principle-Based Theory?

The scholars who conduct purely theoretical studies are not particularly numerous in the modern literature, yet one can find intriguing contributions. For instance, Nozer Singpurwalla discusses the Bayesian methods in relation to various dependability issues [3]. His works provide both rigorous mathematical answers to the raised issues and philosophical insights. Thompson [4], Ascher [5] and Feingold [6] make a neat distinction between repairable and non-reparable systems and underline the consequences of misconceptions on the theoretical and the practical plane as well. Barlow and Proschan put forth reasons for fundamental methodological features including fault tree analysis and accelerated degradation models [7]. Lomnicki explains subtle distinctions such as the difference between the normal 'failure to operate' and the 'failure to idle' that occurs when a system inadvertently does something that should not be done [8]. Steinsaltz and Goldwasser argue on the limits and inherent qualities of the mathematical models used in the study of ageing [9]. There are also philosophers who are concerned with engineering issues such as Harris who criticizes the prediction criteria used in the reliability domain since they provide numbers that are often insignificant and have a poor epistemological basis [10]. A special place is occupied by Gavrilov and Gavrilova [11] who develop a *theory of aging and longevity* employing deductive-logic methods. They will be cited in the next pages more than once.

1.1.1 The scholars who look into purely theoretical topics do not make a large group. They do not raise ample debates in the scientific community despite the high quality of their works. The search for abstract principles does not seem to be an attractive proposition, and this unconcern impels us to address the question:

What is a principle-based theory?

We bring the following historical example case to introduce the theme: principle-based theory.

Example The discovery of fire was one of the earliest of mankind's conquests. Fire's purposes were multiple, some of which were to add light and heat, to cook plants and animals, to burn clay for ceramic objects, to melt metal and create items. Mankind devised a great set of techniques to control fire, but each method was applied apart from the other due to a lack of unifying criteria.

In the 18th century the success of the steam engine triggered the Industrial Revolution; and engineers felt the need for a theoretical aid to perfect and strengthen their engines. Several scientists coming from different countries contributed to set up the edifice of knowledge that we presently call 'thermodynamics'. This engineering discipline enables scholars to derive all that they need simply from the three laws that constitute the kernel of the rigorous principle-based theory.

This achievement was not obtained all at once. James Watt patented his steam engine in 1776 and the first thermodynamic textbook was written in 1859: over eighty years later. During this period partial results saw the light and the entire construction matured as an independent discipline after extensive testing. We believe that the same route is open to reliability studies.

1.1.2 Speaking in general, a principle-based theory deduces all the results from a few initial assumptions; in other words it is grounded in *deductive logic* which proves to be different and somewhat contrary to *inductive logic*, typical of statistical inquiries and very popular in the reliability domain. That is why it is worth recalling some properties of deductive reasoning [12].

A deductive inference starts with the premises $\mathcal{A}$ and draws the conclusions $\mathcal{B}$ using the rules of logic. Each close is certain provided that the argument's assumptions are true. The antecedent logically imposes the consequent

$$\mathcal{A} \Rightarrow \mathcal{B}$$

Specifically a principle-based theory demonstrates logical implications of this type

$$\textit{Principles and Hypotheses} \Rightarrow \textit{Mathematical Statements}$$

The whole construction begins with a few principles or axioms and draws undisputable conclusions from them. Also probabilistic functions, which authors usually adopt in various applications, are to be achieved through mathematical demonstrations. The deductive approach lies far apart from the statistical approach in that the truth of the premises provides a guarantee of the truth of the conclusion. By contrast, using the inductive approach a statistician assigns a likelihood to each conclusion; he seeks the best distribution which fixes or increases the plausibility of the closing statement [13] but cannot ensure that the result is the logical consequence of the premises.

The statistical inference is the process of drawing conclusions about populations or scientific truths from data. There are many modes of performing inference including *statistical modeling* and *data oriented strategies*. An expert has the classic or the Bayesian statistical methods at his disposal. Experts are furnished with a broad assortment of algorithms and likelihood criteria for qualifying the end product of an inductive reasoning process. The premises provide some degree of support for the final statement of an inductive argument but the closing statements never turn out to be absolutely true or false.

Ancient philosophers popularized the term 'deduction' as '*reasoning from the general to the specific*'; and explained 'induction' as '*reasoning from the specific to the general*'. Modern thinkers do not deem these definitions be perfectly accurate, nonetheless they have the virtue of making more evident the difference between the two logical approaches.

1.1.3 The implication $\mathcal{A} \Rightarrow \mathcal{B}$ requires accessory elements and the principle-based theory is usually equipped with:

- A description of the objects at issue which we refer to as *physical model* in the present book,
- A few opening generalizations labeled as *principles* or *axioms*,
- Some *ideal models* that formally describe the behavior or the essential characteristics of the *physical model.*

Literally the term 'principle' means '*a primary or beginning truth from which others are derived*'. The initial statements of entailment necessarily refer to the objects under examination and we call *physical model* the formal description of those objects.

Example The thermodynamic system (TS) is the *physical model* for the devices that exchange matter and energy, and consists of a macroscopic volume determined by its walls in space (Fig. 1.1). A more specific example of TS is gas confined by a piston in a cylinder. The relationships between TS and its surroundings lead to a variety of systems, including *open, closed* and *isolated systems*. The laws of thermodynamics are the primary truths regarding TS and all the thermal studies derive from these initial truths.

The *physical model* is the generalized description of an object and widely differs from the 'statistical model' which gives an account of how one or more random variables are related to other variables [14]. This dissimilarity makes more apparent the distance placed between deductive logic and inductive logic. Perhaps the reader is much more familiar with the latter logic than the former, hence the words *physical model* will be emphasized through Italics in this book.

An important part of the principle-based theory is the so called '*ideal model*', which usually describes an important feature of the *physical model*. The goal of the ideal model is the creation of a typical pattern figure in which the essential elements of the object are included, as well the fundamental features that describe the intended conduct [15].

Examples The reversible process is the *ideal transform* of TS, which can be reversed completely without there being any trace left to show that TS has undergone thermodynamic change. The *ideal circuit* is equipped with electrical elements that have precise and specific reactions; an ideal circuit is not affected—for instance—by eddy currents and thermal fluctuations. The *ideal gas* is composed

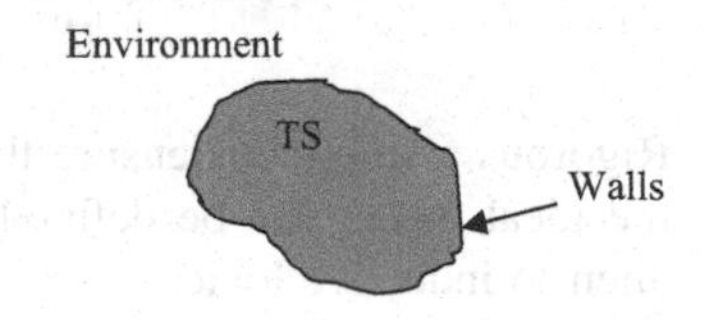

Fig. 1.1 Thermodynamic system

of many randomly moving point particles that do not interact except when they collide elastically. This linear equation governs the ideal gas

$$PV = nRT \tag{1.1}$$

where P is the pressure of the gas; V is the volume of the gas; n is the amount of gas (also known as number of moles); T is the temperature of the gas and R is the molar constant.

1.1.4 From a certain perspective idealization frees the physical model of its imperfections. It can be said that the ideal model $\mathcal{X}$ is the best example of the material event X and turns out to be a useful tool in many applications. In particular it should be emphasized that:

[a] The ideal model is typical of deductive logic in the sense that $\mathcal{X}$ is derived from precise premises;
[b] $\mathcal{X}$ illustrates the properties of X in a very straightforward manner, i.e. (1.1) is a linear equation;
[c] $\mathcal{X}$ is exact in formal terms.

Example Boyle's equation, Charles's and Avogadro's equations—in the order—fix the relationships between two physical properties of a gas while keeping other properties constant

$$\begin{array}{ll} V \propto 1/P, & \text{where } n, T = \text{const.} \\ V \propto T, & \text{where } n, P = \text{const.} \\ V \propto n, & \text{where } P, T = \text{const.} \end{array}$$

These different equations can be combined into a single relationship to make a more general gas law

$$V \propto \frac{nT}{P}.$$

One can introduce the proportionality constant R called *molar* or *universal gas constant* which is featured in many fundamental equations in the physical sciences

$$V = R\left(\frac{nT}{P}\right).$$

And lastly one gets

$$PV = nRT. \tag{1.1}$$

Rigorous assumptions ensure the precision of the ideal model; and for this reason the ideal model can be defined inside a deductive frame whereas it is somewhat alien to inductive logic.

1.1.5 Deductive logic spells out what antecedent $\mathcal{A}$ is responsible for the result $\mathcal{B}$, and therefore the principle-based theory has noteworthy epistemological qualities. It provides scholars with a precise understanding of the world, an understanding of how things happen and why they happen, and the way they do. A principle-based theory widens the vision of practitioners and theorists, and empowers them to forecast future developments in the domain covered by the hypotheses.

The exact implication between the cause $\mathcal{A}$ and the effect $\mathcal{B}$ has no comparison term in statistical inquiries. Also *the inferential studies on the outcomes of causality* achieved by means of the Bayesian methods [16] and *the principle of the common cause* by Reichenbach [17] have nothing to do with deductive logic; instead inductive logic yields probabilistic conclusions.

Further arguments about the 'inductive-statistical' approach in comparison with the 'deductive-nomological' approach may be found in Hempel [18] who explains how the function of deterministic theories goes beyond merely establishing connections between observables. Theories answer the 'why' of things and thus every adequate scientific account is potentially predictive.

1.1.6 A principle-based theory—once it has gained general acceptance in the scientific community—becomes an *exact science* that provides support to experts across various sectors.

$$\textit{Empirical inquiries} \rightarrow \textit{Principle-based theory} \rightarrow \textit{Exact science}.$$

The aid it provides does not vanish with time since the general principles remain valid despite the ever-new technologies that enter the market. For instance, thermodynamics, optics and other formalized disciplines sustained specialists a century ago and do so in present day too. In a similar way, *the reliability science* will provide statements that will be true forever.

1.2 First Cornerstone: Soloviev's Proof

"*Mathematical Methods in Reliability Theory*" [19] illustrates some statistical relationships among the fundamental quantitative characteristics of systems. It also presents methods of finding estimates for reliability parameters based on observations, methods of testing reliability hypotheses and other interesting outcomes. From the perspective of the present book, the contents of the second chapter appear extremely intriguing: *the authors calculate the reliability function and the hazard function through deductive logic*. This special approach and the key parameters obtained qualify this chapter as the first cornerstone to erect reliability as a principle-based construction and subsequently as an exact science. The authors were fully aware of the final target as they wrote: "*It is undeniable that the reliability theory is a complex science.*"

1.2.1 We are going to comment on the demonstration of the notion of reliability that was developed by Alexander Soloviev. We recognize his authorship in the present section, at the same time, we recall how Gnedenko managed and checked all parts of the cited book. Gnedenko is, by any reckoning, one of the most eminent Russian mathematicians and one of the most wide-ranging and influential authors. He is often recognized as the 'father of the reliability theory'. That is why Gnedenko is mentioned as the reference-author in the title of the present book and in the subsequent sections.

1.2.2 Soloviev assumes the *Markov chain* as the suitable *physical model* for a system which works and finally breaks down.

Soloviev observes how each operational unit passes through a series of events; he calls **A** the event during which the elements of the unit function without failure in the interval (0, t), and **B** the event without failure in (t, t_1) (Fig. 1.2). He calculates $P(t, t_1)$—the probability of steady work between t and t_1—using the formula of conditional probability

$$P(t,t_1) = P(\mathbf{A}/\mathbf{B}) = P(\mathbf{AB})/P(\mathbf{A}). \tag{1.2}$$

He makes explicit $P(\mathbf{A})$ as $P(t)$; and $P(\mathbf{AB})$ as $P(t_1)$ because the event **AB** makes steady work in (0, t_1)

$$P(t,t_1) = P(t_1)/P(t). \tag{1.3}$$

The probability of failure in (t, t_1) is

$$Q(t,t_1) = 1 - P(t,t_1) = [P(t) - P(t_1)]/P(t). \tag{1.4}$$

Soloviev poses $t_1 = (t + \Delta t)$ when Δt tends to zero, and obtains

$$Q(t,t+\Delta t) = \frac{P(t) - P(t+\Delta t)}{P(t)} \approx -\frac{P'(t)}{P(t)}\Delta t = \lambda(t)\Delta t. \tag{1.5}$$

The *hazard rate* $\lambda(t)$ (also called *failure* or *mortality rate*) is the instantaneous incidence rate and determines the reliability of the system in each moment this way

$$\lambda(t) = -P'(t)/P(t). \tag{1.6}$$

From (1.5) and (1.6) Soloviev obtains the probability of good functioning which we label as *general exponential function*

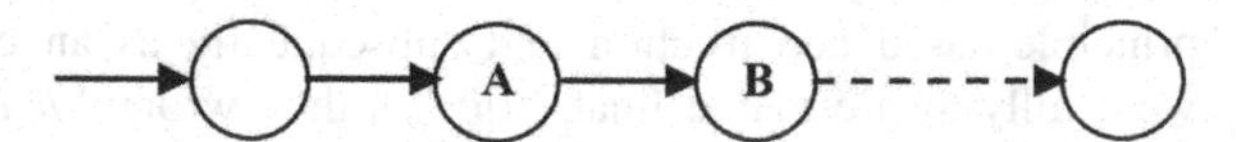

Fig. 1.2 Markov's chain

$$P(t) = \mathrm{e}^{-\int_0^t \lambda(t)dt} \tag{1.7}$$

This function establishes *the concept of reliability* in mathematical terms.

1.3 Comments on the Reliability Function

Soloviev demonstrates the reliability function (1.7) on the basis of the conditional probability which is typical of the Markov chain, and teaches that the sequential occurrence of the system operations results in the general exponential function.

1.3.1 Steps from (1.2) to (1.7) itemize the overall deductive implication which we express as follows

$$\text{Chained Operations} \Rightarrow \boxed{\begin{array}{l} P_f(t) \text{ is general exponential function;} \\ \lambda(t) \text{ is generic.} \end{array}}$$

The Markovian dependence makes comprehensible the physical origin of failures in general.

1.3.2 Section 1.2.2 occupies a small area inside the book [19] but has a great significance in relation to the reliability domain. Equation (1.7) begins to position the notions in their proper places and could be defined as the '*first law of the reliability science*'. Gnedenko, Soloviev and Bélyaev demonstrate how the present sector can become a highly quantitative discipline that derives various technical indices from general statements. Actually, the Russian authors provide the first contribution to the construction of this mathematical 'edifice'.

1.4 Questioning the Bathtub Curve

The hazard function has a generic form in (1.7) and Soleviev adds that λ(t) has three principal trends. He holds that "*much experimental data*" show that a new system has a decreasing failure rate in the early part of its lifetime during which it is undergoing burn-in or debugging. This period is followed by an interval when failures are due to causes resulting in a constant failure rate. The last period of life is one in which the system is experiencing the most severe wear out and the *bathtub curve* (*tripartite* or *U-shaped curve*) has become the hallmark of system reliability

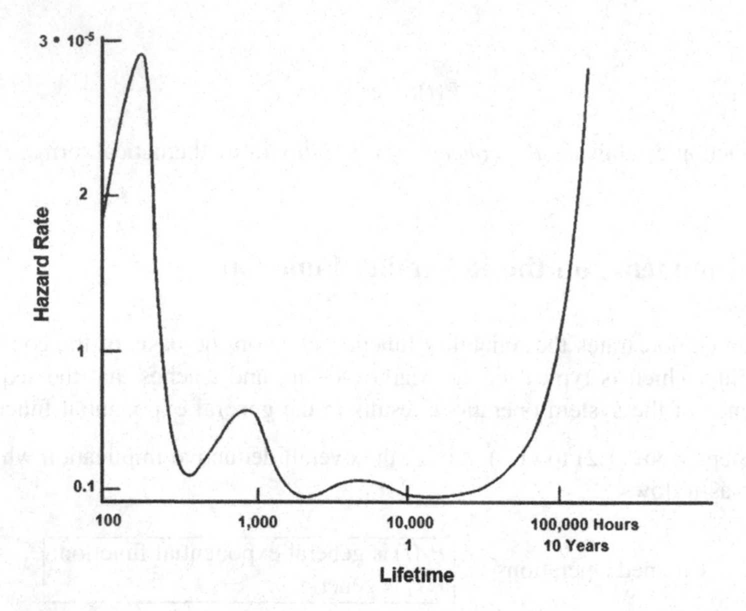

Fig. 1.3 The hazard rate of an electrical circuit. Reproduced from [20] with permission from IEEE

status. This scheme aligns with common sense, but several researchers have criticized it.

The adverse arguments are not negligible since it is not exaggerated to hold that the general exponential function (1.7) centers on the hazard rate. If $\lambda(t)$ is confusing or wrong, then $P(t)$ is also questionable and the mathematical achievements by Soleviev risks vanishing. We can but pay attention to the disapproval expressed by a large segment of the literature.

1.4.1 Several empirical inquiries bring evidence of various irregularities in the hazard rate of devices. For instance, Jensen [21] comments on the ample fluctuations of $\lambda(t)$ in electronic circuits There are such evident humps that Wong labels this a '*roller coaster distribution*' (Fig. 1.3) [22].

The amount of upshots against the bathtub curve is even larger in the biological domain. Aird claims that for humans the infant mortality period is followed by the onset of the wear-out period without delay, such that the constant failure rate period is negligible [23]. The results obtained by Carey and others are based on an initial number of approximately 35,000 medflies (Mediterranean fruit flies) [24]. The mortality of three female groups makes a *concave down curve* (Fig. 1.4) whereas the classic $\lambda(t)$ is a *concave up curve*.

Recently Jones, Scheuerlein and others conducted a large study—published by the journal *Nature* [25]—on the age-patterns of 11 mammals (including humans), 12 other vertebrates, 10 invertebrates, 12 vascular plants, and a green alga. The

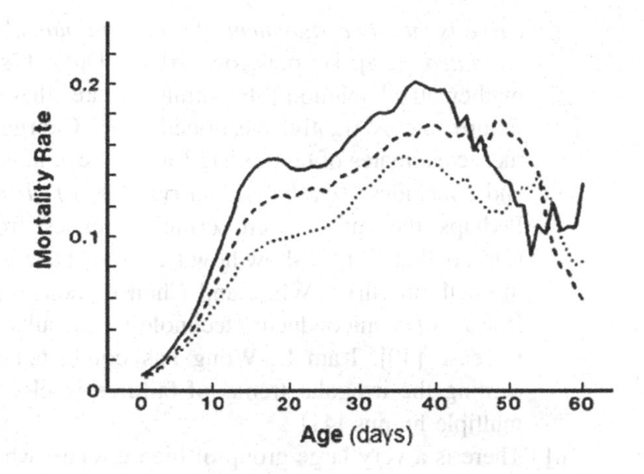

Fig. 1.4 Mortality rates of three cohorts of flies. Reproduced from [24] with permission from Elsevier

mortality of these species exhibits extraordinary variety, including increasing, constant, decreasing, humped and bowed trajectories for both long and short-lived species. The authors conclude: "*This diversity challenges theoreticians to develop broader perspectives on the evolution of ageing*".

1.4.2 Experts react to undisputable negative evidence in various manners. We group the scholars into three circles for the sake of simplicity:

[i] This group expresses a negative judgment about the U-shaped curve in an open manner. Experts claim that the tripartite shape is rarely a physical reality; rather it is often the manifestation of a generic belief. Krohn [26] says that no other justification is given apart from the anthropomorphic parallel with human vitality. Nelson writes: "*a few products show a decreasing failure rate in the early life and an increasing failure rate in later life*" [27]; Marvin Moss holds that: "*the bathtub curve can model the reliability characteristics of a generic piece-part type, but not of an assembly, a circuit, or a system*" [28]. Kececioglu and Sun assert that the bathtub curve describes "*only 10–15%*" of applications [29]; Lynn and Singpurwallaannotate: "*the famous bathtub curve rarely has a physical motivation" and "the main purpose served by burn-in is psychological*" [30]. Prof. Downtown from the University of Birmingham says that the bathtub curve is another fascinating example of the impact folklore has in reliability studies. Klutke and others illustrate some of the limitations of the traditional bathtub curve, and conclude that the value of this curve in characterizing infant mortalities is questionable [31]. Tsang notices that in the airline industry the U-shaped curve is not a universal mode, and adds: "*as much as 89% of all the airline equipment items do not have a noticeable wear-out region throughout their service life*" [32]. Shooman [33] and Carter [34] who widely investigated military equipment make critical remarks about the tripartite hazard rate. Ascher claims: "*the bathtub*

curve is merely a statement of apparent plausibility which has never been validated" [35]. He puts forward the Drenick's theorem as an alternative mathematical solution [36]. Juran declares that the assumption of constant failure rate is rightly questioned [37]. George argues that the nonparametric estimates of failure rate function can reveal some unexpected things and concludes: "*the bathtub curve doesn't always hold water*" [38].

Perhaps the most severe criticism arises from the electronic sector. Campbell et al. [39] show how field data have different characteristics than the bathtub curve. White and Chen demonstrate that increased miniaturization of semiconductor technologies results in a constant failure rate increase [40]. Kam L. Wong has conducted a systematic inquiry concerning the irregular trends of failures in electronic chips which present multiple humps [41].

[ii] There is a very large group of theoreticians who do not reject the bathtub curve in a direct way rather they search for statistical distributions that adhere to empirical data in a more realistic manner. They even aim to depict irregularities and abnormal trends. Nadarajah identifies 17 principal functions currently in use [42].

The Weibull distribution has perhaps become the most popular model in the engineering field because it is flexible enough to model a variety of data sets [43]. The Weibull distribution can depict failures which have decreasing, increasing or constant trends, and allow experts to describe any phase of the item's lifetime. For years, researchers have been developing various extensions of the Weibull distribution, with different numbers of parameters. A bibliographical survey on the generalization of the Weibull distribution can be found in *Pham* and *Lai* [44].

A circle of authors believes that a mixture of distributions could explain the different trends of the hazard rate and even its humps, irregularities etc. [45]. Among the pioneers, Proschan [46] observed the distinguished behavior of mixtures of distributions. Mann, Schafer and Singpurwalla propose a special mixture of Weibull distributions [47].

Other researchers suggest innovative solutions such as Hjorth [48] who proposes a statistical distribution with three parameters obtained by generalizing the Rayleigh function which itself is a generalization of the exponential distribution. Gaver and Acar consider a statistical family that is similar but more complex than Hjorth's distribution which has 4 parameters [49].

Some authors have sought simpler mathematical methods; for example Jones and Hayes [50] who investigate a new reliability-estimation method that does not depend upon current hazard model.

[iii] Lastly, a large group of writers accept the traditional U-shaped curve and disregard any contrary evidence. They appreciate the clear notions that back the tripartite curve and use it to treat an assortment of topics dealing with dependability. We shall return to this topic in Chap. 8.

1.4.3 The above-mentioned groups assume contradictory positions in particular a blatant discrepancy can be observed between [i] and [iii].

The achievements of the first two groups have substance beyond any doubt, but [i] basically relinquishes the idea that a comprehensive framework can provide a full description of the hazard rate. The statistical models devised by the authors [ii] ameliorate the description of $\lambda(t)$ but do not pursue the objective of explaining the physical origins of its differing trends. The failure rate is still one of the most commonly used notions in engineering and in survival analysis but the research strategies of [i] and [ii] do not indicate alternative principle-based schemes to create.

Presently scholars are incapable of understanding the strange behaviors of $\lambda(t)$ through a unified logic. All this has direct impact on the value of Eq. (1.7) that formalizes the concept of reliability $P(t)$ and centers on the hazard rate function. Before $\lambda(t)$ has been perfectly clarified, experts risk becoming confused with respect to this essential aspect of their area of interest.

1.5 Scopes, Methods and Overview of the Present Work

Gnedenko and his colleagues accept the U-shaped curve on a rather intuitive basis and the empirical description of the hazard rate is the weak point inside their theoretical frame. It may be said that they work as mathematicians in the first stage—when deriving function (1.7)—and work as practitioners in the second stage—when they sanction the bathtub curve beyond a precise mathematical proof. Gnedenko, Soloviev and Bélyaev set up a *correct theory* in point of logic since the demonstration from (1.2) to (1.7) is perfect and provides a clear explanation of the system reliability using general hypotheses. Nowadays we should simply search for the proof of $\lambda(t)$ that details its multifold conduct.

1.5.1 Gnedenko was right: *reliability is a new science*. The way that he opened indicates an important destination and the scientific community has never left out his lesson.

Speaking in general, the abstract foundation of a scientific discipline is not something of little importance and normally a group of researchers who intervene at different stages, brings this demanding job to an end. By way of illustration, we recall anew the case of thermodynamics and the many eminent scholars who worked to achieve the same goal. James Joule (1818–1889), Sadi Carnot (1796–1832), Rudolf Clausius (1822–1888), William Thomson (1824–1907) best known as Lord Kelvin, Ludwig Boltzmann (1844–1906), James Maxwell (1831–1879), Willard Gibbs (1839–1903) and others created a huge theoretical edifice over the course of numerous decades. It is natural that also the foundations of the reliability science should share a similar destiny and build on the contributions of various researchers.

1.5.2 We personally decided to investigate the enigmatic trends of the hazard rate function and published partial results during the last decade [51–56]. The present book brings together the various results and aims to provide the comprehensive view of this inquiry; it includes three parts:

- Part 1 describes the formal tools that will be used and their conceptual origin. Next we calculate the hazard rate of functioning systems.
- Part 2 demonstrates the properties of reparable systems.
- Part 3 sums up the main points of the book and comments on the whole work.

This summary shows how we have the aim of proceeding with Gnedenko's project. This book derives the different trends of $\lambda(t)$ from distinct hypotheses that are congruent with the Gnedenko's assumptions; in the subsequent step this book calculates some features of repairable systems.

1.5.3 In the light of Gnedenko's seminal work we shall develop the classic concept of the hazard rate which refers to a system's capability of working, and ignore other views on the hazard rate such as the *hazard potential* which expresses the system resistance to failure [57].

References

1. Ushakov, I. (2012). Reliability theory: History and current state in bibliographies. *Reliability: Theory & Applications, 1*(24), 8–35.
2. Poincaré, H. (1908). *La Science et l'hypothèse*. Paris: Éditeur Ernest Flammarion.
3. Singpurwalla, N. D. (2002). Some cracks in the empire of chance (flaws in the foundations of reliability). *International Statistical Review, 70*(1), 53–78.
4. Thompson, W. A. (1981). On the foundations of reliability. *Technometrics, 23*(1), 1–13.
5. Ascher, H. (2005). Eyeball analysis suffices for important insights ignored by theorists for 50 years. In *Proceedings of the Symposium on Stochastic models in Reliability, Safety, Security and Logistics* (pp. 16–22). Beer-Sheva (Israel): Publisher Sami Shammoon College of Engineering.
6. Ascher, H., & Feingold, H. (1984). *Repairable systems reliability: Modeling, inference, misconceptions and their causes*. New York: CRC Press/Marcel Dekker, Inc.
7. Barlow, R. E., & Proschan, F. (1975). *Statistical theory of reliability and life testing*. New York, N.Y.: Holt, Rinehart and Winston.
8. Lomnicki, Z. A. (1973). Some aspects of the statistical approach to reliability. *Journal of the Royal Statistical Society. Series A, 3*, 395–419.
9. Steinsaltz, D., & Goldwasser, L. (2006). Aging and total quality management: Extending the reliability metaphor for longevity. *Evolutionary Ecological Research, 8*(8), 1445–1459.
10. Harris, L. N. (1985). The rationale of reliability prediction. *Quality and Reliability Engineering International, 1*(2), 77–83.
11. Gavrilov, L. A., & Gavrilova, N. S. (1991). *The biology of life span: A quantitative approach*. New York: Harwood Academic Publisher.
12. Lawson, A. E. (2000). The generality of hypothetic-deductive reasoning: Making scientific thinking explicit. *The American Biology Teacher, 62*(7), 482–495.

13. Kass, R. E. (2011). Statistical inference: The big picture. *Statistical Science, 26*(1), 1–9.
14. Mc Cullagh, P. (2002). What is a statistical model? *The Annals of Statistics, 30*(5), 1225–1310.
15. Dilworth, C. (2007). *The metaphysics of science: An account of modern science in terms of principles, laws and theories* (2nd ed.). Dordrecht: Springer.
16. Williamson, J. (2004). *Bayesian nets and causality: Philosophical and computational foundations*. New York: Oxford University Press.
17. Wronski, L. (2014). *Reichenbach's paradise: Constructing the realm of probabilistic common causes*. Berlin: De Gruyter Open.
18. Hempel, C. (1958). The theoretician's dilemma. In H. Feigl, M. Scriven, & G. Maxwell (Eds.), *Minnesota studies in the philosophy of science* (Vol. II, pp. 37–98). Minneapolis: University of Minnesota Press.
19. Gnedenko, B. V., Bélyaev Y. K., & Solovyev, A. D. (1966). *Математические методы в теории надёжности*. Nauka; Translated as: *Mathematical methods in reliability theory*. Academic Press (1969); and as: *Methodes mathematiques en theorie de la fiabilite*. Mir (1972).
20. Jensen, F. (1989). Component failures based on flaw distributions. In *Proceedings of the IEEE Annual Reliability and Maintainability Symposium* (pp. 91–95).
21. Jensen, F. (1996). *Electronic component reliability*. New York: Wiley.
22. Wong, K. L. (1989). The roller-coaster curve is in. *Quality and Reliability Engineering International, 5*(1), 29–36.
23. Aird, J. (1978). Fertility decline and birth control in the People's Republic of China. *Population and Development Review, 4*(2), 225–253.
24. Carey, J. R., Liedo, P., & Vaupel, J. W. (1995). Mortality dynamics of density in the Mediterranean fruit fly. *Experimental Gerontology, 30*(6), 605–629.
25. Jones, O. R., Scheuerlein, A., Salguero-Gomez, R., Camarda, C. G., Schaible, R., Casper, B. B., et al. (2014). Diversity of ageing across the tree of life. *Nature, 505*(7482), 169–173.
26. Krohn, C. A. (1969). Hazard versus renewal rate of electronic items. *IEEE Transactions on Reliability, 18*, 64–73.
27. Nelson, W. (1982). *Applied life data analysis*. New York: Wiley.
28. Moss, M. A. (1985). *Designing for minimal maintenance expense: The practical application of reliability and maintainability*. New York: Marcel Dekker.
29. Kececioglu, D., & Sun, F. (1995). *Environmental stress screening: Its quantification, optimization, and management*. Prentice Hall.
30. Lynn, N. J., & Singpurwalla, N. D. (1997). Comment: 'Burn-in' makes us feel good. In H. W. Block & T. H. Savits (Eds.), *Burn-in, ser. Statistical Science* (Vol. 12, pp. 1–19).
31. Klutke, G. A., Kiessler, P. C., & Wortman, M. A. (2003). A critical look at the bathtub curve. *IEEE Transactions on Reliability, 52*(1), 125–129.
32. Tsang, A. (1995). Condition-based maintenance: tools and decision making. *Journal of Quality in Maintenance Engineering, 1*(3), 3–17.
33. Shooman, M. L. (1968). *Probabilistic reliability: An engineering approach*. New York: McGraw-Hill.
34. Carter, A. D. S. (1973). *Mechanical reliability*. Macmillan.
35. Ascher, H. (1968). Evaluation of repairable system reliability using the "bad-as-old" concept. *IEEE Transactions on Reliability, 17*(2), 103–110.
36. Drenick, R. F. (1961). The failure law of complex equipment. *Journal of the Society for Industrial and Applied Mathematics, 8*(4), 680–690.
37. Juran, J. M. (1988). *Juran's quality control handbook* (4th ed.). New York: McGraw-Hill.
38. George, L. (1994). The bathtub curve doesn't always hold water. *Reliability Review, 14*(3), 5–7.
39. Campbell, D., Hayes, J., Jones, J., & Schwarzenberger, A. (1992). Reliability behavior of electronic components as a function of time. *Quality and Reliability Engineering International, 8*(3), 161–166.

40. White, M., & Chen, Y. (2008). *Scaled CMOS technology reliability user guide*. Pasadena, CA: Jet Propulsion Laboratory, California Institute of Technology. JPL Publication 08–14 3/08.
41. Wong, K. L. (1988). Reliability growth: Myth or mess. *IEEE Transactions on Reliability, 37* (2), 209.
42. Nadarajah, S. (2009). Bathtub-shaped failure rate functions. *Quality & Quantity, 43,* 855–863.
43. Lai, C. D., & Xie, M. (2006). *Stochastic ageing and dependence for reliability*. New York: Springer.
44. Pham, H., & Lai, C. H. (2007). On recent generalization of the Weibull distribution. *IEEE Transactions on Reliability, 56,* 454–458.
45. Wondmagegnehu, E. T., Navarro, J., & Hernandez, P. J. (2005). Bathtub shaped failure rates from mixtures: A practical point of view. *IEEE Transactions on Reliability, 54*(2), 270–275.
46. Proschan, F. (1963). Theoretical explanation of observed decreasing failure rate. *Technometrics, 5,* 373–383.
47. Mann, N. R., Schafer, R. E., & Singpurwalla, N. D. (1974). *Methods for statistical analysis of reliability and life data*. New York: Wiley.
48. Hjorth, U. (1980). A reliability distribution with increasing, decreasing, constant and bathtub-shaped failure rates. *Technometrics, 22*(1), 99–107.
49. Gaver, D. P., & Acar, M. (1979). Analytical hazard representations for use in reliability, mortality, and simulation studies. *Communications in Statistics B, 8,* 91–111.
50. Jones, J., & Hayes, J. (2001). Estimation of system reliability using a "non-constant failure rate" model. *IEEE Transactions on Reliability, 50*(3), 286–288.
51. Rocchi, P. (2000). System reliability and repairability. In *Livre des Actes de la Deuxième Conférence Internationale sur les Méthodes Mathématiques en Fiabilité* (Vol. 2, pp. 900–903). Bordeaux: Publisher Lab. Statistique Math, et ses Applications.
52. Rocchi, P. (2002). Boltzmann-like entropy in reliability theory. *Entropy, 4,* 142–150.
53. Rocchi, P. (2006). Calculations of system aging through the stochastic entropy. *Entropy, 8,* 134–142.
54. Rocchi, P. (2010). A contribution to the interpretation of the late-life mortality deceleration. In *Proceedings of the International Symposium on Stochastic Models in Reliability Engineering, Life Science and Operations Management* (pp. 932–937). Beer Sheva (Israel): Publisher Sami Shammoon College of Engineering.
55. Rocchi, P., & Tsitsiashvili, G. S. (2012). Four common properties of repairable systems calculated with the Boltzmann-like entropy. *Applied Mathematics, 3*(12A), 2026–2031.
56. Rocchi, P. (2015). Can the reliability theory become a science? *Reliability: Theory and Application, 10*(36), 84–90.
57. Singpurwalla, N. D. (2006). The hazard potential: Introduction and overview. *Journal of the American Statistical Association, 101*(476), 1705–1717.

Part I
Functioning Systems

Chapter 2
Old and Novel Tools for the Calculus of the Hazard Rate

The major subjects covered in this chapter revolve around the following topics:

- Deductive reasoning needs a precise description of the intended objects of study and we begin with discussing the *physical models* of systems.
- The concept of entropy is very prolific in science [1] and this chapter will put forward the Boltzmann-like entropy as an innovative mathematical tool to calculate the systems' hazard rate.

We recall that Shannon's entropy is used in a broad variety of research areas but here we shall introduce a new form of entropy that has nothing to do with the concepts of information and noise. Secondly, we shall overlook software systems.

2.1 The Reliability of What?

A principle-based theory should calculate artificial and natural systems, man-made and biological entities and, in fact, Gnedenko et al. [2] adopts the *chain* as the *physical model* capable of describing this multitude.

2.1.1 We mean to particularize the hazard rate depending on different circumstances where the Markovian chain assumes a variety of forms, hence we take that S is a *continuous-time stochastic system with a set of discrete states* and append the following remarks:

1. A device or a biological entity normally has some alternative macro-states, i.e. a body is living or otherwise dead; a machine is running or otherwise is under reparation. The *m states* or *macro-states* are mutually exclusive, and we obtain the structure of S

P. Rocchi, *Reliability is a New Science*, DOI 10.1007/978-3-319-57472-1_2

$$S = (A_1 \, OR \, A_2 \, OR \ldots OR \, A_m), \quad m > 0. \tag{2.1}$$

2. Each state is equipped with *sub-states* (sometimes called *components* or *parts*) which work together toward the same purpose. That is to say, the generic state A_i includes *n sub-states* that cooperate in order to fulfill the mission of A_i

$$A_i = (A_{i1} \, AND \, A_{i2} \, AND \ldots AND \, A_{in}), \quad i = 1, \ldots m; \, n > 1. \tag{2.2}$$

3. We integrate (2.1) and (2.2) and obtain the *structure of levels,* which has three grades of granularity at least. We give the name *macro-scale level, meso-scale level* and *micro-scale level* to them from the top to the bottom. This can be also called an *OR/AND structure* because of the relationships that govern it

$$\begin{aligned} S &= (A_1 \, OR \, A_2 \, OR \ldots OR \, A_r) \\ &= (A_{11} \, AND \, A_{12} \, AND \ldots) OR \ldots OR (A_{r1} \, AND \, A_{r2} \, AND \ldots). \end{aligned} \tag{2.3}$$

A stochastic system is basically a Markov chain however we shall not exploit the Markovian dependencies; rather we shall look into the various forms of the structure of level and the interrelationships amongst the components. It may be said that we shall examine the *topology of S*.

2.1.2 Structure (2.3) can be placed close to the engineering design schemes which represent a device at different scales of measurement. The levels allow the progressive zoom of S and offer a basic aid to exploring intricate systems. Factually, the levels prove to be flexible in analyzing failures that includes different mechanisms at different metric scales.

Example Biologists hold that the human body encompasses the immune system, the circulatory and respiratory systems, and many others. Each system includes organs e.g. the circulatory system is equipped with the heart, veins and arteries. Each organ is composed of tissues and the tissues are made of cells. The structure of the human body includes six principal levels

$$\begin{aligned} \textit{Human Body} &= \{\text{Alive}\} \, OR \, \{\text{Dead}\} \\ &= \{[\text{System}_A] \, AND \, [\text{System}_B] \, AND \, [\text{System}_C] \, AND \ldots\} \, OR \, \{\ldots\} \\ &= \{[(\text{Organ}_{A1}) \, AND \, (\text{Organ}_{A2}) \, AND \ldots] \, AND \, [\ldots] \ldots\} \, OR \, \{\ldots\} \\ &= \{[(\text{Tissue}_{A11} \, AND \, \text{Tissue}_{A12} \, AND \ldots) \, AND \, (\text{Tissue}_{A21} \, AND \ldots) \ldots] \, AND \, [\ldots] \ldots\} \, OR \, \{\ldots\} \\ &= \{[((\text{Cell}_{A111} \, AND \, \text{Cell}_{A112} \, AND \ldots) \, AND \, (\text{Cell}_{A121} \, AND \, \text{Cell}_{A122} AND \ldots) \ldots) \ldots] \, AND \, [\ldots] \ldots\} \, OR \, \{\ldots\} \end{aligned} \tag{2.4}$$

2.1.3 The hierarchical decomposition of systems is rooted in the pioneering works of Herbert Simon [3] who writes:

> Scientific knowledge is organized in levels, not because reduction in principle is impossible, but because nature is organized in levels, and the pattern at each level is most clearly discerned by abstracting from the detail of the levels far below.

The *hierarchy theory* offers efficient suggestions on how to surmount complexity in nature. The structure serves to subdivide a complex system into organizational levels and into discrete components within each level. The hierarchy theory clarifies other concepts such as surfaces, stability, nesting, filters and the integration of disturbances into systems [4].

2.1.4 States and sub-states are mapped into probabilities which indicate how likely it is that S stays in those states and sub-states

$$P_i = P(A_i).$$

One can calculate the probability in accordance with the relationships *OR* and *AND*. In particular, the *AND* amongst the sub-states in (2.2) provides the probability of A_i

$$P_i = (P_{i1} \cdot P_{i2} \cdot P_{i3} \cdot \ldots \cdot P_{in}), \quad n > 1. \tag{2.5}$$

2.1.5 In principle, [5] the essential states of S in the reliability domain are the following:

A_f = the *functioning state* during which the system runs,
A_r = the *recovery state* during which the system is repaired, or renewed and so forth,
A_l = the *idle state* during which the system is good and idles,
A_d = the *dereliction state* during which the broken system is not manipulated.

Thus (2.1) has this precise form

$$S = \left(A_f \, OR \, A_r \, OR \, A_l \, OR \, A_d\right). \tag{2.6}$$

The system evolves from one state to another according to the transition graph in Fig. 2.1.

2.1.6 We shall investigate various stochastic structures that vary depending on the goal pursued by each enquiry. We shall tailor the structure of levels according to the problem being tackled, specifically:

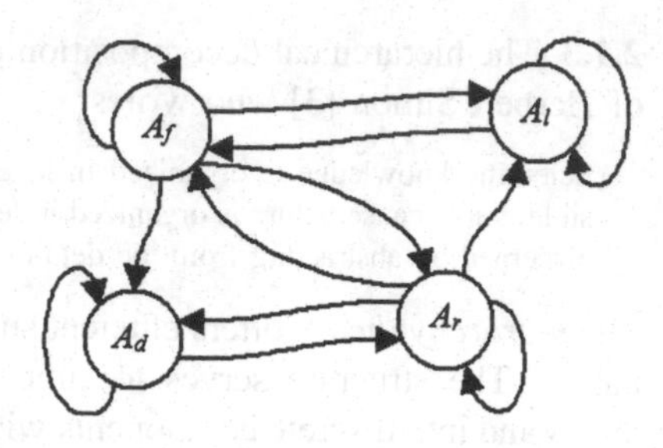

Fig. 2.1 System states and transactions

- The present chapter discusses the general properties of systems using this *physical model*

$$S = (A_1\, OR\, A_2\, OR \ldots OR\, A_m), \quad m > 0. \tag{2.1}$$

$$A_i = (A_{i1}\, AND\, A_{i2}\, AND \ldots AND\, A_{in}), \quad i = 1, 2, \ldots m;\ n > 1. \tag{2.2}$$

- Chapters from 3 to 8 examine exclusively the functioning state

$$S = A_f. \tag{2.7}$$

- Chapter 9 studies the dynamical system that is repaired or maintained

$$S = \left(A_f\, OR\, A_r\right). \tag{2.8}$$

2.1.7 Machines, devices and other artificial constructions are built as determined by specified standard. Manufacturers scrap the units that do not pass the quality control phase. This management method allows theorists to employ the stochastic model in engineering; but living beings diverge from man-made artifacts. Generically speaking, biological systems do not undergo any 'quality control' and are more intricate than the artificial. Individuals of the same species differ one another, and even present monstrous deformities at the birth.

Are mathematical models, specifically stochastic models, appropriate *physical models* for living beings?

All mathematical models simplify the reality, but they can offer some significant advantages: a relatively simple explanation can depict complex situations. A rigorous pattern can predict experimental data with precision. As second we recall how different studies require different viewpoints of the intended facts, and depending on the purpose scientists develop different models for the same phenomenon. The present work means to clarify the various aspects of the mortality rate function, and theorists in the reliability domain show how the stochastic model can add a great deal to our understanding of living beings' life span and longevity [6].

2.2 Entropic Systems

Generically speaking, the use of the entropy function is not new in the treatment of failures but the entropy function which we are going to present, the issues that we mean to tackle and the direction which we address are very different from the approaches taken in the current literature. Despite some trivial similarities we have followed a special pathway; that is why we discuss the argument since its early beginnings.

2.2.1 When we started to undertake the fundamental issues of reliability, we believed that new mathematical tools should be devised and thermodynamics proved to be a fertile ground for inspiration.

Experts in thermodynamics noted how a certain amount of energy released from combustion is always lost inside a thermal engine and thus is not transformed into useful work. Uncompensated transformations led theorists to define the *reversible process* which basically can be restored to the initial state from the final state without requiring a modification of the properties of the surrounding environment. By contrast, the *irreversible process* is a thermodynamic change which can be reversed thanks to an additional exchange of energy with the context (Fig. 2.2).

The entropy H_T qualifies the thermodynamic cycle T which starts from the state A—we tacitly refer to equilibrium states—and goes back to it

$$T = (A \rightarrow B \rightarrow A)$$

In particular Clausius' inequality holds that for the reversible process T the change in entropy is zero. Whereas the change in the entropy H_T is always lower than zero for the irreversible cycle

$$\Delta H_T = \oint \frac{\delta Q}{T} \leq 0. \tag{2.9}$$

Where δQ is the heat exchanged at any point of the cycle T and T the temperature of those exchange points. Note how the reversible system is an ideal model, whereas all the physical systems are irreversible in the reality.

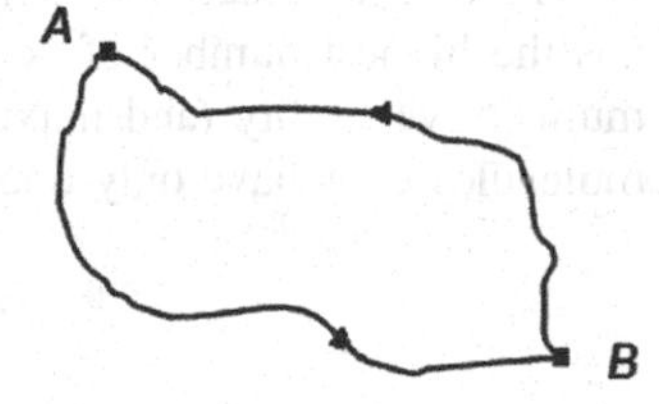

Fig. 2.2 The process T

2.2.2 Ludwig Boltzmann devised the thermodynamic entropy H_A that has two special properties:

- It calculates the property of the thermodynamic states A, B, C etc. and not the process T,
- It is related to the notion of probability and not to heat Q and temperature T, the parameters typical of H_T in (2.9).

Factually, the state A of the thermodynamic system (*TS*)—e.g. a sample of gas—is realized by different positions and momenta of the molecules contained in that sample. The energy of *TS* can be arranged in different ways and each *complexion of molecules* corresponding to a single A constitutes a *microstate* of A. The thermodynamic state A has several complexions or microstates: a, b, c, d....

$$A = (a\, OR\, b\, OR\, c\, OR \ldots).$$

Boltzmann defines his entropy this way

$$H_A(W) = k_B \log(W_A). \tag{2.10}$$

Where W_A is the number of alternative complexions which correspond to the macroscopic state A and k_B is Boltzmann's constant. The entropy $H_A(W)$ is the natural logarithm of the number of possible microscopic configurations of the individual molecules which give rise to the observed macroscopic state A of *TS*.

The thermodynamic system can assume only one microstate at a time, hence the number of favorable cases is 1 and the total number of possible cases is W_A. The various complexions are equally likely and the probability of the generic complexion X is

$$P_X = 1/W_A.$$

It seems useful to specify these details because there is a certain confusion between the so-called *thermodynamic probability* W_A—which Boltzmann introduces but is not the authentic mathematical probability—and the mathematical probability P_X [7].

2.2.3 Boltzmann notices that the higher is W_A and the higher the irreversibility of A. The thermodynamic system spontaneously reaches the *equilibrium state* which has the highest number of possible microstates W_A. On the other hand, the minimum irreversibility (and maximum reversibility) occurs when W_A is the unit; the molecules of A have only one microscopic configuration, and the entropy is null

$$H_A(W) = k_B \log(1) = 0. \tag{2.11}$$

This situation happens at the absolute temperature zero

$$T = T_0 = 0\,^{\circ}\mathrm{K}. \tag{2.12}$$

All the molecules are motionless and this entails that there is only one mechanical microstate which pertains to the macroscopic T_0-state. Various theorems demonstrate that it is impossible by any material procedure—no matter how idealized—to reduce the temperature of any system to T_0 in a finite number of operations. The T_0-state is so much reversible that a physical device can merely approximate it.

In Boltzmann's view, reversibility is the feature opposite to irreversibility, in the sense that when one increases, the other diminishes and vice versa. This double way relation could be written as follow

$$\textit{Reversibility}/\textit{Irreversibility} \tag{2.13}$$

We shall cite the coupled properties R and I with the symbol: R/I.

2.2.4 It is demonstrated that phenomena manageable in the real world are irreversible, including gas heating, liquid diffusion, friction between solid surfaces and chemical reactions. As a consequence, the *closed* thermodynamic system cannot finish a real process with as much useful energy as it had to start with. Some energy is always wasted. The second law of thermodynamics holds that the entropy of *TS* —either H_T or H_A—becomes greater in the world and is often enunciated in this form

$$\textit{The entropy of the universe tends to a maximum.} \tag{2.14}$$

Entropy can decrease with time in systems that are not closed. Factually, many *TS*s reduce local entropy at the expense of an environmental increase, resulting in a net increase in entropy.

From one point of view, statement (2.14) holds that the physical reality has an inherent tendency towards disorder and a general predisposition towards decay. A huge process of annihilation has been going on everywhere in the universe according to classic thermodynamics, and this suggested to us that there was an intriguing parallel with the reliability theory when we started to investigate the general trend of $\lambda(t)$. We observed how thermal processes and reliable systems alike are condemned to evolve toward demolition and these parallel behaviors intimated to us to introduce the entropy function for the study of reliable/reparable systems *S*. We had the intention to express some general properties of *S* in a manner similar to the properties of *TS* and introduced a function symmetric to (2.10) to which we gave the name *Boltzmann-like entropy*.

2.3 Boltzmann-like Entropy

We apply the notion of reversibility and irreversibility to the *physical model* (2.1).

2.3.1 For the sake of simplicity, suppose that there are two alternative and rather extreme situations for the stochastic system *S*. Once it has entered the generic state A_i ($i = 1, 2, \ldots m$), it may happen that *S* runs steady in A_i or otherwise *S* moves from A_i. In the former case *S* does not evolve from A_i, this state is rather stable and we say that A_i is *irreversible*. In the latter event, *S* abandons A_i, this state is somewhat unstable and we say A_i is *reversible*. E.g. a man/woman goes into an irreversible coma, this means that he/she does not leave this state and no longer recovers. E.g. the machine *S* was immediately repaired, this means that the failure was minor and the failure state was very reversible. In substance, the notion of reversibility is not simply a matter of 'playing the movie' backwards; the joined properties reversibility/irreversibility are a construct that allows one to judge the quality of the system behavior as it is being performed as well as evaluate its consequences.

2.3.2 We intend to calculate R/I of the state A_i in similarity with the Boltzmann entropy. Mathematicians prove the function $H_A(W)$ on the basis of three axioms [8] which we intend to examine:

(1) The number of complexions *W* expresses a certain idea of frequency; in parallel we demand that the Boltzmann-like entropy of the state A_i depends on the probability P_i.
(2) The entropy $H_A(W)$ is minimal when A is absolutely reversible [see (2.11)]; the Boltzmann-like entropy likewise reaches the minimum when A_i is *reversible*. The Boltzmann-like entropy is at its maximum when the stochastic state is *irreversible*.
(3) The microstates of a gas sample influence the overall state A and in turn the entropy H_A. Similarly we assume that the irreversibility of each component A_{i1}, A_{i2}, ... A_{in} affects the state A_i. By way of illustration, suppose that the part A_{ik} reaches an irreversible state because it breaks down, and affects the whole machine which stops. In practice, the entropy of A_{ik} changes the entropy of the overall operational state A_i and in turn the whole equipment. As a further example, let all the parts of a machine work steadily and in consequence the system fairly runs. The irreversible functioning micro-states A_{f1}, A_{f2}, ... result in the irreversible functioning macro-state A_f. The present assumption, which relates the R/I of the parts to the R/I of the whole, is the most meaningful one. It clarifies the system's complexity and the behavior of its parts.

We translate criteria (1), (2) and (3) into three assumptions as follows.

2.3.3 Let the generic state A_i has n cooperating parts

$$A_i = (A_{i1} AND\, A_{i2} AND. \ldots AND\, A_{in}), \quad n > 1. \tag{2.2}$$

(1) The Boltzmann-like entropy H_i of A_i is continuous in the P_i

$$H_i = H(P_i). \tag{2.15}$$

(2) $H(P_i)$ is a monotonic increasing function of P_i

(3) $H(P_i)$ is an additive function, namely it is the summation of the entropies of the sub-states A_{i1}, A_{i2},... A_{in}

$$H_i = H_{i1} + H_{i2} + H_{i3} + \ldots + H_{in}. \tag{2.16}$$

Theorem 2.1 *The H_i satisfying assumptions (1), (2) and (3) is of the form*

$$H_i = H(P_i) = a\log(P_i) + b, \quad a > 0,\ b \geq 0. \tag{2.17}$$

Where a is the scale parameter, and b is the position parameter that depend on the system. We give the name of Boltzmann-like entropy to H_i.

Proof In order to simplify the formal expressions, let the state A_i consist of two sub-states (n = 2). We apply the multiplication law from (2.5)

$$P_i = P_{i1} \cdot P_{i2}. \tag{2.18}$$

From axiom (3) we obtain

$$H = H(P_{i1} \cdot P_{i2}) = H(P_{i1}) + H(P_{i2}). \tag{2.19}$$

We develop this expression by the Taylor series and preliminary we unify the variables in this way

$$\begin{aligned} P_{i1} &= P_i, \\ P_{i2} &= 1 - \varepsilon. \end{aligned} \tag{2.20}$$

Where ε is infinitesimal of the first order. Formula (2.19) becomes

$$H(P_i - \varepsilon P_i) = H(P_i) + H(1 - \varepsilon).$$

We develop this expression in series and neglect the terms of the highest order

$$H(P_i) - \varepsilon P_i \cdot H'(P_i) = H(P_i) + H(1) - \varepsilon \cdot H'(1). \tag{2.21}$$

The maximum of the entropy function is zero in accordance with (2.11)

$$H(1) = 0. \tag{2.22}$$

Hence

$$P_i \cdot H'(P_i) = H'(1). \tag{2.23}$$

Namely

$$P_i \cdot H'(P_i) = k, \quad k \geq 0. \tag{2.24}$$

This expression means

$$H'(P_i) = \frac{k}{P_i}. \tag{2.25}$$

The integration yields

$$H(P_i) = a \log(P_i) + b. \tag{2.26}$$

For the sake of simplicity we use $a = 1$ and $b = 0$ (Fig. 2.3), and obtain

$$H(P_i) = \log(P_i). \tag{2.27}$$

Secondly, the Boltzmann entropy depends on the number of the molecules complexions W_A that varies between 1 and $+\infty$. The domain of $H_A(W)$ is $(0, +\infty)$, instead $H(P_i)$ varies within

$$(-\infty, 0). \tag{2.28}$$

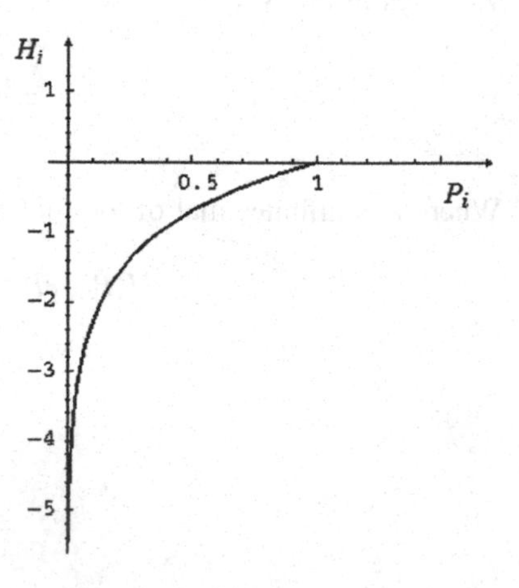

Fig. 2.3 The Boltzmann-like entropy

2.4 Physical Meaning of the Boltzmann-like Entropy

The Boltzmann-like entropy is intended to study the evolvability of this system

$$S = \left(A_f \, OR \, A_r \, OR \, A_l \, OR \, A_d\right). \tag{2.6}$$

The idle and the dereliction states do not carry out any transformation for S and do not contribute to any change. They turn out to be inessential for the entropic study, we confine attention to A_f and A_r, and in turn we inquiry:

(1) The *reliability or functional entropy* $H_f = H(A_f) = H(P_f)$
(2) The *recovery entropy* $H_r = H(A_r) = H(P_r)$

2.4.1 H_f and H_r specify the inherent characteristic of S in the functioning state and the recovery state in the order. The following intuitive remarks can help the reader understand 'how' the *functional entropy* H_f and the *recovery entropy* H_r depict S in the states A_f and A_r.

(1.a) When H_f is 'high', the functioning state is irreversible and the system works steadily. In particular, the more H_f is high, the more A_f is irreversible, and S is *capable of working*.
(1.b) On the other hand, when H_f is low, S often abandons A_f and switches to H_r, we say that *S is incapable of working*.
(2.a) When H_r is 'high', the recovery state is irreversible, and the workers operate on S with effort. In particular, the more H_r is high, the more A_r is stable, and in practice *S is hard to repair* and/or maintain in the world.
(2.b) On the other hand, when H_r is low, S leaves A_r and one says that *S can be easily restored or maintained.*

I/R are coupled opposite characters in accordance to (2.13); thus the function H_f illustrates the *capability to work* (point 1.a) and also the *incapability of working* (point 1.b). The function H_r qualifies the *capacity to be repaired* (point 2.b) and the *incapacity of being repaired* (point 2.a). The following two statements merge annotations 1.a with 1.b; and 2.a with 2.b as follows:

The reliability entropy expresses the aptitude of S to work without failures;
The recovery entropy illustrates the disposition of S toward reparation. (2.29)

2.4.2 Practitioners adopt a very high number of variables to calculate the devices' capability of working including maximum speed, minimum noise, rapid answers etc. They assess several parameters even within a single engineering context. Doctors measure up the health state of a patient using many clinical tests. The vigor of a young body is qualified by means of fatigue resistance, indifference to low/high environmental temperature, low blood pressure etc. Some theorists qualify the performances of S using a set of states; for example, Mesbah et al. [9] introduce a four-state model which includes the *good, neutral, bad* and *dead states*. This scale supports longitudinal inquiries of the quality of life. In summary, many criteria are employed to qualify the effectiveness of each class of systems; it seems noteworthy that the entropy H_f expresses the aptitude of S to work which does not depend on the technology, on the form and dimension of S, on the nature of S—say artificial or natural—and on other special features.

Usually engineers qualify the labor required to repair S using an assortment of markers: the number of work-hours, the replaced components, the financial costs and so forth. Doctors give an account of the days necessary for a patient to recover from a surgery, an infection or another disease, and the various treatments required: drugs, hours of physiotherapy etc. The recovery entropy H_r expresses the aptitude of S to be repaired by reason of the reversibility and irreversibility criterion that crosses numerous fields of application.

2.4.3 Let us examine a pair of numeric examples.

Example Suppose a and b are two devices in series with $P_f(a) = 10^{-200}$, $P_f(b) = 10^{-150}$. We can calculate the probability of good functioning and then the stability of the overall system with the entropy

$$P_f(S) = [P_f(a) \cdot P_f(b)] = 10^{-350}$$

$$H_f(S) = \log\left[P_f(S)\right] = \log(10^{-350}) = -805.9$$

The R/I of the whole S depends on the R/I of the parts, the Boltzmann-like entropy is additive [assumption (2.16)] and one can follow this way with the same result

$$\begin{aligned} H_f(S) &= \left[H_f(a) + H_f(b)\right] = \log\left[P_f(a)\right] + \log\left[P_f(b)\right] \\ &= -345.3 + (-460.5) \approx -805.9 \end{aligned}$$

Example Suppose a device degrades during the interval (t_1, t_2); and the probability values are the following: $P_f(t_1) = 10^{-10}$, $P_f(t_2) = 10^{-200}$. The entropies qualify the irreversibility of the device

$$H_f(t_1) = \log\left(10^{-10}\right) = -23.0$$

$$H_f(t_2) = -460.5$$

One obtains how much the capability of good functioning has slowed down

$$\Delta H_f = H_f(t_2) - H_f(t_1) = -460.5 - (-23.0) = -437.5.$$

2.4.4 The calculation of the Boltzmann-like entropy does not oppose great difficulties in numerical terms; instead it raises somewhat serious arguments from the conceptual viewpoint as the significance of the free variable of $H(P)$—the probability P—is still arguable from many aspects.

The *frequency* and the *subjective theories* are currently under discussion and in a way these constructions underpin the classical and Bayesian statistics respectively. Some results of classical statistics have a similar Bayesian counterpart, and this has compounded the question even more. The correspondences between the two statistical schools have encouraged some working statisticians to minimize the differences and emphasize how parallel results can be obtained in both the statistical courses [10, 11].

Someone still objects that the mentioned differences and even other divergences do not seem dramatic, they appear smaller in magnitude compared to the distances between the parametric versus nonparametric approaches, the design versus model modes etc.

We mean to answer as follows.

The debate about the divergences of the classical and Bayesian statistics and even the discussion on the superiority of one approach over the other, remains on the surface of the questions until one assumes an operational viewpoint. As matter of facts, nobody has a meter in hand that is capable of measuring the virtues of a statistical method. There is no shared scale to quantify the distance between the right side and the left side of Table 2.1.

We note how the substantial contrast between the classical and Bayesian statistics does not emerge during the calculation phase but when the parallel use of the two statistical methods finishes. We mean to say that a statistician gets two numerical values at the end of the calculations, but the probability obtained through the classical method is *an objective quantity*, the probability obtained through Bayesian procedures expresses *a personal belief*. The two probabilities can even be equal in numerical terms but their meanings turn out to be irreconcilable (Table 2.2).

Table 2.1 Some opposite traits of the statistical methods

Classical Statistics	Bayesian Statistics
Unknown variables are treated as deterministic quantities that happen to be unknown	Unknown variables are treated as random variables with known distributions
Classical statisticians concentrate on the full set of data that might arise	Bayesians concentrate on the observed data only
The existence of true parameters is assumed	True parameters are not supposed to exist
Underlying parameters remain constant during repeated sampling: parameters are fixed	Parameters are unknown and described probabilistically: data are fixed

Table 2.2 Opposite conclusions of the two statistics

Classical Statistics	Bayesian Statistics
Probability is a physical and objective quantity	Probability is a means to quantify uncertainty

In consequence of the disparate significance of P, the Boltzmann-like entropy $H(P)$ assumes altogether contrasting significance and the purpose of this book, which means to provide a unified ground to the dependability domain, seems destined to be short lived.

The recent volume [12] addresses the problem of the probability interpretation and remarks that the debates about the double nature of probability has apparent philosophical origin. The book presents a pair of theorems to circumvent the philosophical approach that unfits mathematical logic. The theorems demonstrate that a statistician can select either the classic or the Bayesian approach without logical incongruence. As a result, a scholar becomes aware of what kind of entropies are involved whenever probability is defined in terms of chances, i.e. physical probabilities, or credence, i.e. degrees of belief.

2.4.5 A non-negligible group of scholar shares the idea that there is a certain parallel between thermodynamics and reliability. Feinberg and Widom [13] write this sentence which we fully subscribe to:

> The reliability science for physics-of-failure lacks a basic foundation. Thermodynamics is a natural candidate. Many engineers do not realize how closely thermodynamics is tied to reliability.

Some authors view degradation as an irreversible process that dissipates energies and can be qualified by the entropy function. They extend thermodynamics to the sector of dependability in a direct manner and calculate the variation of total thermodynamic entropy dH in order to assess the damage of components. Experimentalists apply this method especially in the mechanical context [14–16]. This vein of research, which could be labeled '*thermodynamic reliability*,' is summarized by the following cause-effect inference

$$\textit{Degradation} \rightarrow \textit{Damage} \rightarrow \textit{Dissipated energy} \rightarrow \textit{Entropy production}$$

In a way, this cause-effect analysis mirrors the initial motivation of the present work since both the approaches read the entropy function as an optimal tool that is capable of qualifying degradation and does not depend on the choice of observables. Although, there are great differences in the purposes and the mathematical methodologies between the present approach and the approach of the thermodynamic reliability, the state-of-the-art of which is illustrated in the work of Amiri and Modarres [17]. The latter deems degradation to be lost energy, whereas the former sees degradation as a somewhat reversible state of functioning. The latter calculates the thermodynamic entropy; the former the Boltzmann-like entropy. The present approach aims to tackle the definition of the hazard rate in general, researchers in thermodynamic reliability address specific empirical cases.

2.4.6 The *dynamical system theory* is a significant theoretical area that deals with the long-term behavior of systems and their evolution over time. The basic elements of the dynamical system theory are:

- The set X of all possible *states* x of the dynamical system.
- The *phase space* $\mathbb{R}^n$ of the system; $x \in \mathbb{R}^n$.
- The *family of transformations* T_t: $X \rightarrow X$ mapping the phase space to itself.

The transformations depend on time notably when t is a positive real number, the system's dynamics is continuous. If t is a natural number, S execute a discrete set of actions [18]. The Kolmogorov-Sinai (KS) entropy measures the difficulty to foresee S, in the sense that the higher the unpredictability is, the higher the KS entropy. This property matches nicely with the Shannon entropy, where the unpredictability of the next character coming from the source is equivalent to new information. It is also consistent with the concept of entropy in thermodynamics, where disorder increases the entropy; in fact, disorder and unpredictability are closely related. The KS entropy provides a significant aid to understanding the complexity of dynamical systems [19]. For instance, it contributes to the *chaos theory*, a mathematical frame of deterministic behaviors that are characterized by great disorder and confusion.

The dynamical system theory presents some traits that are not so distant from the present frame. For instance, Courbage and Prigogine [20] examine the concept of the irreversibility of dynamical systems. However, the objectives of that theory differ from the scope of the present work, and the pathways that the two constructions follow are diverging.

References

1. Frigg, R., & Werndl, C. (2010). Entropy: A guide for the perplexed. In C. Beisbart & S. Hartmann (Eds.), *Probabilities in physics*. New York: Oxford University Press.
2. Gnedenko, B. V., Bélyaev, Y. K., & Solovyev, A. D. (1966). *Математические методы в теории надёжности*. Nauka; Translated as: *Mathematical methods in reliability theory*. Academic Press (1969); and as: *Methodes mathematiques en theorie de la fiabilite*. Mir (1972).
3. Simon, H. S. (1973). The organization of complex systems. In H. H. Pattee (Ed.), *Hierarchy theory: The challenge of complex systems* (pp. 1–27). New York: George Braziller.
4. Wu, J. G. (2013). Hierarchy theory: An overview. In R. Rozzi, J. B. Callicott, S. T. A. Pickett, & J. J. Armesto (Eds.), *Linking ecology and ethics for a changing World: Values, philosophy, and action* (pp. 115–142). Cambridge Press.
5. Ushakov, I. A. (2012). *Probabilistic reliability models*. Hoboken: Wiley.
6. Steinsaltz, D., Mohan, G., & Kolb, M. (2012). Markov models of aging: Theory and practice. *Experimental Gerontology, 47*(10), 792–802.
7. Lavenda, B. H. (1991). *Statistical physics: A probabilistic approach*. New York: Wiley.
8. Chakrabarti, C. G., & Chakrabarty, I. (2007). Boltzmann entropy: Probability and information. *Romanian Journal of Physics, 52*(5–6), 525–528.
9. Mesbah, M., Dupuy, J. F., Heutte, N., & Awad, L. (2004). Joint analysis of longitudinal quality of life and survival processes. In N. Balakrishnan & C. R. Rao (Eds.), *Advances in Survival Analysis* (Vol. 23, pp. 689–728). Elsevier.
10. Kim, Y., & Schmidt, P. (2000). A review and empirical comparison of Bayesian and classical approaches to inference on efficiency levels in stochastic frontier models with panel data. *Journal of Productivity Analysis, 14*, 91–118.

11. Berger, J., Boukai, B., & Wang, Y. (1999). Simultaneous Bayesian–frequentist sequential testing of nested hypotheses. *Biometrika, 86,* 79–92.
12. Rocchi, P. (2014). *Double-faced probability*. Springer.
13. Feinberg, A. A., & Widom, A. (2000). On thermodynamic reliability engineering. *IEEE Transactions on Reliability, 49*(2), 136–146.
14. Klamecki, B. E. (1984). An entropy based model of plastic deformation energy dissipation in sliding. *Wear, 96,* 319–329.
15. Doelling, K. L., Ling, F. F., Bryant, M. D., & Heilman, B. P. (2000). An experimental study of the correlation between wear and entropy flow in machinery components. *Journal of Applied Physics, 88,* 2999–3003.
16. Naderi, M., Amiri, M., & Khonsari, M. (2010). On the thermodynamic entropy of fatigue fracture. *Proceedings of the Royal Society, 466,* 423–438.
17. Amiri, M., & Modarres, M. (2014). An entropy-based damage characterization. *Entropy, 16,* 6434–6463.
18. Downarowicz, T. (2011). *Entropy in dynamical systems*. New York: Cambridge University Press.
19. Schuster, H. G. (1995). *Deterministic chaos: An introduction*. Wiley.
20. Courbage, M., & Prigogine, I. (1983). Intrinsic randomness and intrinsic irreversibility in classical dynamical systems. *Proceedings of the National Academy Science USA, 80,* 2412–2416.

Chapter 3
How Systems Break Down

We intend to formulate a principle-based theory through implications of this kind

$$\mathcal{A} \Rightarrow \mathcal{B}$$

Hence we are obliged to begin with the study of the premises $\mathcal{A}$ which are necessary for deriving the results.

3.1 Interferences Overlap

This initial part of the book centers on the functioning state

$$S = A_f. \tag{2.7}$$

Thus we can use A_f and S as synonyms.

3.1.1 In accordance with the current literature we investigate S that operates until the first interruption. By 'interruption' we mean the break-down or cessation of the normal operations of a device or of a living being. Hence, the probability P_f (t) coincides with the *probability of functioning with no failure as time goes by* and equals $P(t)$ calculated with (1.7).

3.1.2 As matter of facts, agents which cause work interruptions rarely work alone; a number of failure determinants often act together and put the system on the path to failure. The premises $\mathcal{A}$ of a failure are rarely simple and this universal property is to be fixed by means of the following principle:

***Principle of Superposition**: Failure factors of any kind usually overlap and shape the hazard rate.*

P. Rocchi, *Reliability is a New Science*, DOI 10.1007/978-3-319-57472-1_3

This basic assumption establishes that multiple causes of failure are to be taken into account, and the contribution of various interfering elements enormously complicates the study of $\lambda(t)$.

3.1.3 The cumulative contribution of various determinants to a single failure is often obscure as long as they are hidden and even work asynchronously. James Reason set up the *latent failure theory* [1] that explains how a failure agent may play an *active* role or may be *silent* for a while. For instance a human error—such as the procedure violation during a surgery—provokes damage that does not emerge in that moment but remains hidden for a while. A latent defective part may lie dormant for days, weeks, or months until it contributes to the definitive system's end. Reason puts forward an excellent visual model known as the *Swiss cheese model* that explains when some different factors can contribute to a failure and when they cannot do so (Fig. 3.1).

This model likens system's components to multiple slices of Swiss cheese, stacked side by side. The holes of the slices represent the hidden flaws which are located in various parts of the system. Considering the holes of a slice as opportunities for a process to fail, an error may allow a problem to pass through a hole in one layer, but in the next layer the holes are in different places, and the problem is caught. For a catastrophic failure to occur, the holes need to align at each step.

The Swiss cheese model shows how different mechanisms placed in different points and occurring in different times contribute to the system's end. It demonstrates how the *principle of superposition* seems simple on paper but turns out to be intricate in the world.

3.1.4 Several studies investigate how multiple mechanisms result in a single failure. For example Gloeckler [2], Hagen [3], Katz and Konstam [4] analyze the *common cause of failure* of some special events.

3.1.5 In statistics a *mixture distribution* describes random variables that are drawn from more than one parent population [5]. The mixture models are among the most statistically mature methods for clustering and provide a natural representation of heterogeneous elements [6]. The researchers in reliability theory who develop the mixture models share, in a way, the inclusive perspective adopted in the present book.

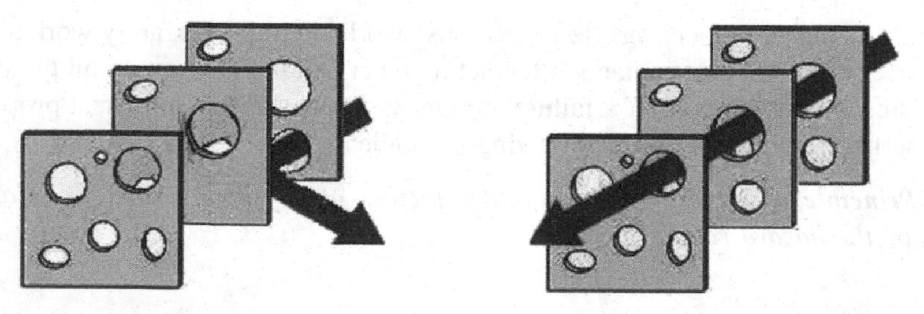

Fig. 3.1 Swiss cheese model with potential failure (*left*), and actual failure (*right*)

3.2 Factors Leading to Work Interruption

It is not easy to classify the elements $\mathcal{A}$ which lead to the work interruption and we adopt the *continuity/discontinuity criterion* that is amply shared in literature; read for example Federal Standard 1037C [7]. The continuity/discontinuity criterion subdivides the agents by their temporal duration and establishes the following groups:

1. ***Continuous factors (CF)*** operate without interruption during the system lifetime until *S* breaks down. The system wears out and eventually stops.
2. ***Random factors (RF)*** intervene all at once at a certain point during the system lifetime. Actually, they cannot be foreseen and lead the system to stop functioning.

This classification appears to be rather exhaustive since determinants 1 and 2 are mutually exclusive and a third way is not allowed. We are going to discuss point 1 below and point 2 in Chap. 5.

3.3 Systematic Factors of Failure

Normally engineers optimize a device within the requirements and constraints set by various sources: customers, the environment, materials etc. Regardless of the ability of designers and manufacturers a man-made solution—say a mechanical engine or electric equipment—will have various unwanted effects which progressively injure the components and lead the artifact to cease functioning. Dangerous factors also operate inside living beings and progressively worsen their performances and cause death. Hence we focus on the *continuous side effects* (CSEs) that are the adverse physical mechanisms—no matter they are internal or external or both—that hinder the functioning parts of *S* during its lifetime. Continuous side effects are the most significant continuous factors (CF); CSEs are the universal root-causes of systems' *wear and tear* and we plunge into them.

3.3.1 Collateral effects have various natures: mechanical, electrical, thermal, chemical and many others. Sometimes a CSE and the affected element share the same nature; alternatively, the side effect does not belong to the technical domain of the endangered part; for example, unwanted oxidations corrode an electrical cable: chemical reactions and electrical technology pertain to distinct physical domains. A single CSE causes dissimilar and even contrary responses by components of the same species. For instance, suppose a series of mechanical loads act upon a steel spring at regular intervals, that spring loses elasticity as time passes. When the same sequel of loads acts on a rigid steel girder—that is a mechanical element—eventually the girder bends. One can conclude that the steel girder has become pliable whereas the steel spring has come to be less malleable.

3.3.2 Collateral effects involve huge or tiny parts; they may be clearly evident or even imperceptible; they operate at any tier of the structure of levels (2.3); they ceaselessly operate or may even stop and start at intervals. CSEs are active when *S* runs and may be in action even when *S* is idle. For instance, the quality of objects made with rubber spontaneously gets worse over time even if they are not employed.

3.3.3 The destructive property of continuous side effects is that they can be mitigated but cannot be discarded; nobody can wipe out a continuous side effect. Designers and manufacturers usually minimize the impact of a CSE but are incapable of eliminating it. For example, lubricant oil reduces attrition in a petrol engine but cannot prevent the erosion of surfaces and the wasting of materials. The engine progressively loses power and will cease functioning after a certain number of working hours.

A side effect can interfere in a systematic manner with a component, as one cannot observe that component without interference. A side agent is as bound to a certain phenomenon as it hides the properties of that phenomenon and hinders its study and measurement. For example, a body will spontaneously continue in its state of uniform motion in a straight line, but this natural conduct rarely occurs in the world. The attrition of the soil, the air or other elements systematically contrasts the moving body. Ancient scientists were incapable of conceiving frictionless motion and put forward bizarre theories to interpret the properties of moving bodies. Newton's laws of mechanics shed light on the issues raised by the false impression caused by the ubiquitous attrition.

In conclusion, there is a broad variety of CSEs that although share a common property: collateral effects cannot be avoided and cannot be discarded.

Let us look at some cases of side effects by way of illustration.

Example Suppose a chemical process begins with the reactants *A* and *B*; and brings forth the chemical products *C* and *D*. One is inclined to conclude that the system has the shape on the left side in Fig. 3.2; in reality this scheme is rather misleading since any chemical process exhibits an equilibrium between the forward and the backward reactions. This concept was first developed by Claude Louis Berthollet who established the reversibility of the chemical reactions using the *stoichiometric coefficients a, b, c* and *d* of the respective reactants and products

$$aA + bB \rightleftarrows cC + dD.$$

The *law of mass action* specifies that there is a constant proportion between the forward and backward rates

Fig. 3.2 Idealized chemical process (*left*) and real chemical process (*right*)

$$k_+\{A\}^a\{B\}^b = k_-\{C\}^c\{D\}^d.$$

Where $\{A\}$, $\{B\}$, $\{C\}$ and $\{D\}$ are the active masses of the chemical elements, and k_+ and k_- are the rate constants. At equilibrium one obtains the equilibrium constant K_S which is typical of the intended process S

$$K_S = \frac{k_+}{k_-} = \frac{\{C\}^c\{D\}^d}{\{A\}^a\{B\}^b}.$$

The law of mass is sufficient to illustrate how chemical systems are affected by reverse reactions that diminish the effectiveness of the direct reaction (Fig. 3.2 right). The enduring presence of the reactants is a form of intrinsic pollution that damages the outcomes of the chemical process [8].

Example Several physical CSEs exist in semiconductor devices including *mobility, thermal heating* and *tunneling* which result in the device breaking down [9]. By way of illustration, mobility can be understood as the relocation of material or charge. Atoms, ions, electrons, or holes are shifted from their designated place to a harmful position. The mobility is influenced by the lattice of the circuit and its thermal vibrations, impurity atoms, surfaces and interfaces, interface charges and traps, the carriers themselves, the energy of the carriers, the lattice defects and other collateral effects. An isolated single movement usually does not cause a device failure. However, in highly down-scaled semiconductor devices, a single defect can also lead to device failure. The accumulation around an initial single defect leads to further degradation of the device until the parameters shift out of their specifications, and a severe failure occurs.

Example A great challenge in biology is to analyze the forces shaping biological decline. A variety of contributions have been put forward for analyzing senescence and the number of ageing theories seems to exceed 300. Scholars agree on placing the theories into two major boxes labeled *wear and tear aging theories* and *programmed aging theories* [10]. The former center on certain kinds of side agents which systematically cause damages to the human body such as chemical toxins, metal ions, free-radicals, hydrolysis, glycation, *cross-linking* and *disulfide bond.* Authors also investigate systematic environmental determinants, such as nutrition, access to water, and lacking shelter from excessive temperature [11]. Programmed aging is determined by special components that are responsible for the organism's control—such as longevity genes, the immune system influence, and DNA elements and force elderliness and deterioration.

The wear and tear aging theories fit perfectly with the concept of CSE. The programmed aging theories will be commented on in Chap. 7.

3.4 Simplified Models and Realistic Models

The previous examples illustrate some collateral effects, and even show how they overlap in accordance to the *principle of superposition*. The intricate ensemble of mechanisms that degrade S should be accurately illustrated

$$\begin{aligned} S &= A_f \\ &= (A_{f1}\ AND\ A_{f2}\ AND \ldots AND\ A_{fn}). \end{aligned} \tag{3.1}$$

3.4.1 Theorists often spell out the generic sub-state A_{fg} (g = 1,2,...n) with the 3-tuple

$$A_{fg} = (X_g, Y_g;\ l_g). \tag{3.2}$$

where the pair X_g and Y_g symbolizes the input/output or the initial/final conditions of the operation l_g. Normally a graph visualizes A_{fg} as in Fig. 3.3. These algebraic and graphical schemes turn out to be very manageable but furnish a simplified account since CSEs have been omitted. The complete illustration of A_{fg} should contain the k collateral effects which create hindrance to the operations

$$A_{fg} = \left[(X_g, Y_g;\ l_g); (X_{g1}, Y_{g1};\ l_{g1}), (X_{g2}, Y_{g2};\ l_{g2}), \ldots (X_{gk}, Y_{gk};\ l_{gk})\right], \quad k > 0. \tag{3.3}$$

The graph that complies with 3.3 includes l_g and the multiple failure factors that work at the same time (Fig. 3.4).

The present study spurns the simplified model of S since the CSEs are missing and argues on the *realistic model* of systems.

Example Fig. 3.2 (right side) illustrates a practical case coming from the chemical sector.

Example Carnot defines the ideal model for a heat engine that includes two bodies at temperature T' and T'' ($T' > T''$), the gas h that does the mechanical work L via cycles of contractions and expansions (Fig. 3.5 left).

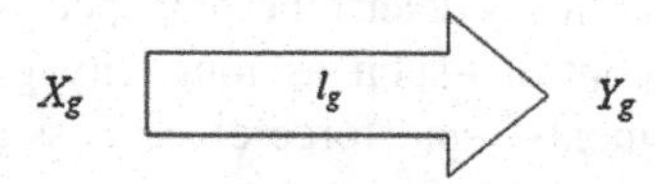

Fig. 3.3 Simplified model of the sub-state A_{fg}

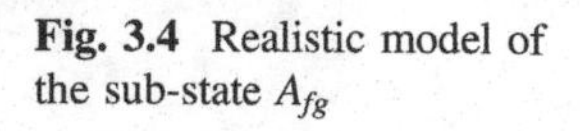
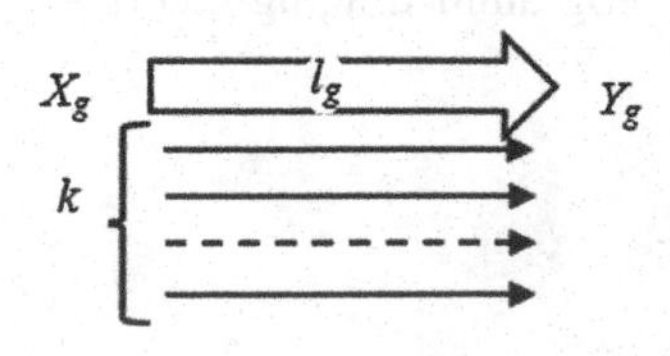

Fig. 3.4 Realistic model of the sub-state A_{fg}

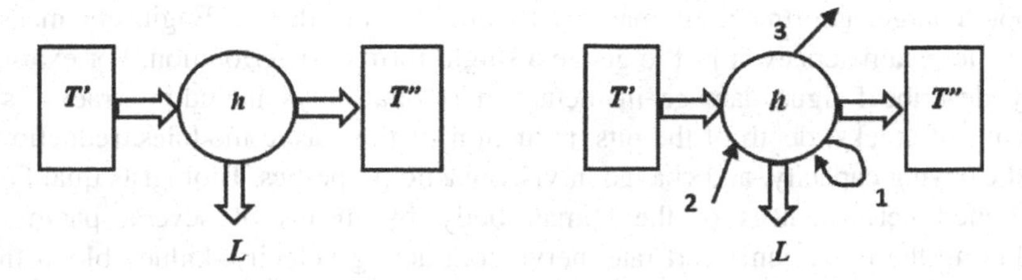

Fig. 3.5 The ideal model (*left*) and the realistic model (*right*) of a heat engine

The realistic model of the heat engine should also visualize the collateral mechanisms such as the heat dispersion (3), the friction amongst the gears (2) and the mounting disorder of the molecules (1) (Fig. 3.5 right). This last agent can be measured by means of Clausius' inequality, which gives the negative change in entropy in a thermal cycle

$$\Delta H = \oint \frac{\delta Q}{T} < 0. \quad (2.9)$$

Excess entropy is created in any phase of the real process. More heat is dumped in the cold reservoir to get rid of this entropy and this leaves less energy to do work.

3.4.2 We put (3.3) into (3.1) in order to examine what CSEs bring out

$$\begin{aligned} S &= A_f(A_{f1}\,AND \ldots A_{fg} \ldots) \\ &= [((X_1, Y_1; l_1); (X_{11}, Y_{11}; l_{11}), (X_{12}, Y_{12}; l_{12}), \ldots (X_{1k}, Y_{1k}; l_{1k}))\; AND \\ &\quad \ldots ((X_g, Y_g; l_g); (X_{g1}, Y_{g1}; l_{g1}), (X_{g2}, Y_{g2}; l_{g2}), \ldots (X_{gk}, Y_{gk}; l_{gk})) \ldots], \\ &\qquad g = 2, \ldots n,\; k > 0. \end{aligned} \quad (3.4)$$

Equation (3.4) depicts a somewhat intricate structure of levels lastingly influenced by unwanted intrinsic and extrinsic factors. One cannot detail the lower level of (3.4). Nobody can say whether the lower components are in *OR* or in *AND*, since CSEs usually work in an unpredictable manner. We can merely observe that the lower level affects the upper level and causes the stop of system operations by time passing.

3.5 Enduring Obtrusion and Decadence

We cannot develop the analytical calculus of (3.4) due to the impossibility of analyzing CSEs and we take a different direction.

3.5.1 Authors use the terms '*wear and tear*' to sum up the outcome produced by the intricate actions that decrease in value an asset as a consequence of use and adopt a large assortment of markers to qualify this decay. Engineers measure multiple parameters even just to gauge a single form of deterioration. For example, they describe fatigue damage in metals in several ways including cracks' size, density of cracks, depth of the pits, reduction of the elastic modules, reduction of load carrying capacity, and change in viscoplastic properties. Biologists qualify the worsened performances of the human body by means of several parameters including the maximum heart rate, nerve conducting velocity, kidney blood flow, maximum breathing capacity, maximum work rate and fertility. It may be said that engineers and doctors address specific issues at the lower level of (3.4). The variables typical of a sector, are suitable within a narrow scope but are not appropriate when they come to defining the conduct of the systems in general. The specialist approach is unsuited to the multifold nature of failure mechanisms.

At this point, we recall Sect. 2.4 which explains how the reliability entropy qualifies the *capability of running;* and in particular, the entropy drop-off describes the progressive decline of the system performances. The function $H_f(t)$ has various virtues; it has the property of being indifferent to the various damaging factors, the paths to failure, the quality of materials, and so on. It applies to spontaneous and man-made structures, to natural and artificial systems, to small and large systems. In conclusion, it seems reasonable to select the reliability entropy $H_{fg}(t)$ to establish the systematic deterioration of the sub-state A_{fg} by time passing.

Principle of Progressive Deterioration: *Let t_0 and t are two times belonging to the system lifetime*

$$H_{fg}(t) < H_{fg}(t_0), \quad t > t_0. \tag{3.5}$$

Enduring damaging agents make worse the working capacity of every part of S and $H_{fg}(t)$ diminishes continuously as time goes by. What a part does now it will not do forever; its performances slope is down and will eventually lead the system S to stop functioning. The literature broadly accepts as true that devices and living organisms systematically wear and tear, and lastly die.

3.6 Comments on the Principle of Progressive Deterioration

From the present viewpoint, the concept of entropy promises to combine intricate failure mechanisms, which appear extremely uneven, into a whole. The *principle of progressive deterioration* is consistent with the *principle of superposition* and offers an integrate perspective on the conduct of systems as:

- H_{fg} applies to all the factors which wear out S.
- H_{fg} applies to any component of S.
- H_{fg} is irrespective of the system technology, nature and size.

It may be concluded that the *principle of progressive deterioration* dominates the entire reliability field.

3.6.1 The present frame tends to place all the agents leading the system toward degeneration into a whole and matches with recent studies such as the *network theory of aging* (NTA) that mean to unify the ageing effects from multiple levels of the living organization. The idea that multifold connected processes contribute to the biological senescence is central to NTA. Kowald and Kirkwood [12, 13] put forward the first model of this theoretical perspective. More recently Trubitsyn [14] develops a frame that unites various existing perspectives on aging and is supported by data accumulated in different fields of biology. The author tries to show that the fundamental phenomena that accompany the aging process include an assortment of degeneration actions running in parallel.

Other authors expand further the range of inquiries and correlate the degeneration of micro elements—say atoms and molecules—with environmental macro-agents such as ecosystems and social groups [15].

3.6.2 Ultimately, the principle of progressive deterioration and the general significance of the Boltzmann-like entropy open a new vision of the physical world which needs to be introduced.

Fundamental laws of physics have no preference for a direction in time. If the direction of time were to reverse, the theoretical statement that describes a law would remain true. However, some natural processes prove to be not time-reversible and contribute greatly to understanding the apparent asymmetry of nature in time, despite nature's perfect symmetry in space [16].

The problem of the direction of time has its early source in the debates between Boltzmann and some of his contemporaries. In 1928, the British astronomer Arthur Eddington [17] introduced the expression '*arrow of time*' to represent the one-way property of time due to the entropy that forces events to move toward in one particular direction. He recognizes that molecular motions are intrinsically reversible, but that they tend to lose their organization and become increasingly shuffled with time. Thus the second law of thermodynamics yields the effect termed as the '*thermodynamic arrow of time*'.

Scientists have classified various time-asymmetric processes [18], such as the following.

- *The radiation arrow of time*: Some physical systems carry energy through harmonic oscillators from place to place. The process is called radiation, and the moving patterns of oscillations are known as waves. When the system delivers energy from a single point, energy radiates away as concentric outgoing waves. The opposite processes, in which energy arrives at one point in the form of incoming concentric waves, do not occur in nature.
- *The cosmological arrow of time*: Modern measurements indicate the Big Bang occurred approximately 13.8 billion years ago. The universe has continued to grow bigger and bigger since its beginning, hence it has been argued that time points in the direction of the universe's expansion. This phenomenon became

apparent towards the beginning of the 20th century, thanks to the work of Edwin Hubble and others who established that space is indeed expanding, and that the galaxies are moving ever further apart.

- *The quantum arrow of time*: Significant results in quantum mechanics are Schrödinger's equation and the collapse of the wave associated to a particle which appears to be a time-asymmetric phenomenon. The location of a particle is described by a wave function, which essentially gives the various probabilities that the particle is in one of the many different possible positions. The wave collapses when factually one achieves a measures the particle and this physical phenomenon cannot occur in the inverse direction; it is not time-reversible and creates a special arrow of time.

Philosophers have found broad processes which are asymmetrical in time though they do not depend on physical laws in strict terms. Thinkers argue about:

- *The historical arrow of time*: Layzer [19] notes how history, whether of the living species or of inanimate matter, moves from the simple to the complex, from the undifferentiated to the increasingly differentiated, from states exhibiting little or no ordering to states that manifest higher degree of order and information. For example, the fusion of nuclei in a star starts from hydrogen and helium, which have the simplest atoms, and provides a variety of complex molecules—metals, noble gasses, halogens etc.—as byproducts of the thermonuclear fusion. Humans groups evolve thanks to the help of individual memory and social memory, and assume ever more complex behaviors. The historical arrow of time points towards mounting complexity.
- *The causal arrow of time*: By definition, a cause precedes its outcome. Although it is surprisingly difficult to satisfactorily define *cause* and *effect*, the concept is clear in the events of our everyday lives. By causing something to happen, one is to some extent controlling the future, whereas whatever one might be able to do, one cannot change or control the past. Sometimes the evolution *cause* → *effect* is probabilistic in nature, nonetheless it is considered to be such a rigid rule that it determines the arrow of time.
- *The perceptual arrow of time*: Everyone has the sense that personal perception is a continuous movement from the known—located in the past—to the unknown—located in the future. The unknown-future world constitutes the physical place where one always seems to be moving towards together with the desires, dreams and hopes that are ahead of the human being. The associations (behind = past) and (ahead = future) are culturally determined in themselves. Another side of the psychological passage of time is in the realm of volition and action. People plan and often execute actions intended to affect the course of events in the future, while it is impossible to change past events.

3.6.3 An interesting aspect of the *principle of progressive deterioration* is the special light it sheds on the concept of time direction and the evolution of machines and living beings. The principle states that the capability of working declines over time and never spontaneously reverts. The damages suffered by a system can be minimized by experts but the root-causes of the damages cannot be eradicated and the wear-out of the system can but progress as time goes by. This process, which is irreversible over time, can be labeled as

The deterioration arrow of time.

And it completes the previous list of processes that change in one direction only with time.

3.6.4 The *deterioration arrow of time* turns out to be rather close to the *thermodynamic arrow of time* since both are expressed using the entropy function. The first arrow basically focuses on the structure of *S* that is qualified by the Boltzmann-like entropy $H_f(t)$; the second focuses on the energy involved in the operations of the thermodynamic system qualified by the thermodynamic entropy H_T. Both the arrows establish the decadence of systems. The latter claims that the entropy of *TS* can increase but not decrease, hence some energy is degraded in its ability to do work. The former principle holds that any animate and inanimate system decays and inevitably stops functioning.

3.6.5 In abstract, all the arrows of time commented on above have an influence on people lives but some of them—e.g. the quantum arrow—have so feeble an impact that they appear to be negligible. Those phenomena draw the attention of scholars who are specialized in those arguments and even raise debates, but are ignored by laymen.

The *deterioration arrow of time* is something apart. It conveys a clear message about the ultimate destiny of individuals, which has always attracted man's attention. The physical and mental decline that ends with death constitutes a valuable issue for everybody and all men and women have thoughts and feelings about it: educated and unlettered; coeval and ancient, kings, tycoons and poor. It does not seem exaggerated to conclude that the *deterioration arrow of time* is of concern to people living in any part of the world, and involves more participation on the part of every body than all the remaining time-irreversible phenomena.

3.6.6 When a scientific principle illustrates a universal truth, it normally finds an echo in philosophical works both before and after the principle has been established. The *principle of progressive deterioration* and the *deterioration arrow of time* underpin arguments and topics which have engaged the minds of people since time immemorial and will continue to do so in the future. An enormous amount of books, articles and essays discuss the unremitting decline of objects and living beings as an undisputable and sad reality.

The earliest reflections about the eclipse of the universe may be found in Plato (about 425—347 BC). The *theory of forms* that lies at the heart of his philosophy revolves around the *transiency of the world.* For Plato any item placed around us degenerates; it erodes and breaks down one way or another [20]. He holds that everything comes into existence and passes away in this world, everything is imperfect and decays because of the contingency of the universe, and nothing ever just permanently is. It is not exaggerated to conclude that Plato and his followers put forth the philosophical *principle of progressive deterioration* about two thousand and five hundred years ago.

Another ancient thinker to mention is the Roman writer Lucius Annaeus Seneca (c. 4 B.C.–A.D. 65) who occupies a central place in the literature on Stoicism. He reasoned about human destiny and wrote some telling pages such as the ensuing passage which has a clear connection with the message conveyed by the *principle of progressive deterioration*:

> We do not suddenly fall into death but progress toward it bit by bit. We die each day, since each day a part of life is taken away, and even as we are growing, our life decreases. We pose our infancy, our childhood, then our youth. Whatever time has passed up until yesterday perished. This every day which we are living, we divide it with death [21].

The decline of animate and inanimate entities attracted the attention of eminent thinker including René Descartes, Georg Wilhelm Leibniz, David Hume and Immanuel Kant. More recently William Lane Craig goes back to the contingency of the universe in "*The cosmological argument from Plato to Leibniz*" [22] and makes original annotations.

Present day philosophers still explore human debasement [23, 24] and wonder:

Is human death simply an instance of organic and functional end? ultimately a matter of biology?
If not, on what basis should it be conceived?
Does death have an essence that entails necessary and even jointly sufficient conditions?

Death emerges as an inexplicable conundrum closely related to the problem of human consciousness and in turn to the definition of a person's essence on the understanding that this essence requires the capacity of a cognitive mind. The mystery of the human departure appears to be a so puzzling topic that people appeal to God. Some find satisfactory answers only from the transcendental vision of the world.

There is an enormous amount of works that consider the end of humans also in the scientific literature. Dozens of theories have been devised to clarify the causes of aging. Sometime the researchers give the impression of working in the hope of circumventing the ineluctable destiny of people. Commentators seem to feel a desire to find a remedy that ensures eternal life. The dream of discovering the *elixir of immortality* fascinated ancient alchemists and modern scientists too.

References

1. Reason, J. (1990). *Human error*. Cambridge: Cambridge University Press.
2. Gloeckler, M., Jenkins, C. R., & Sites, J. R. (2003). Explanation of light/dark superposition failure in CIGS solar cells, *Proceedings of the Symposium of the Material Research Society* (Vol. 763, pp. B5.20.1–B5.20.6).
3. Hagen, E. W. (1980). Common-mode/common-cause failure: A review. *Annals of Nuclear Energy, 7*(9), 509–517.

4. Katz, A. M., & Konstam M. A. (2009). *Heart failure: Pathophysiology, molecular biology, and clinical management* (2nd ed.). Netherlands: Wolters Kluwer.
5. Lindsay, B. (1995). Mixture models: Theory, geometry and applications. *NSF-CBMS Regional Conference Series in Probability and Statistics, 5,* 1–163.
6. Stephens, M. (2000). Bayesian analysis of mixture models with an unknown number of components: An alternative to reversible jump methods. *Annals of Statistics, 28,* 40–74.
7. Information Administration. (1997). *Federal Standard 1037C or Telecommunications Glossary of Telecommunications Terms*. Lanham: Government Institutes. Available at: http://public.eblib.com/EBLPublic/PublicView.do?ptiID=1385085
8. Elnashaie, S. S. E. H., & Garhyan, P. (2003). *Conservation equations and modeling of chemical and biochemical processes*. New York: Marcel Dekker.
9. Bazu M., & Bajenescu T. (2011). *Failure analysis: A practical guide for manufacturers of electronic components and systems*. New Jersey: Wiley.
10. Medveev, Z. A. (1990). An attempt at a rational classification of theories of ageing. *Biological Reviews, 65*(3), 375–398.
11. Conn, P. M. (ed.) (2006). *Handbook of models for human aging*. Ameterdam: Elsevier Academic Press.
12. Kowald, A., & Kirkwood, T. B. (1994). Towards a network theory of ageing: A model combining the free radical theory and the protein error theory. *Journal of Theoretical Biology,* 168, 75–94.
13. Kowald, A., & Kirkwood, T. B. (1997). Network theory of aging. *Experimental Gerontology, 32*(4–5), 395–399.
14. Trubitsyn, A. G. (2013). The joined aging theory. *Advances in Gerontology,* 3(3), 155–172.
15. Simkó, G. I., Gyurkó, D., Veres, D. V., Nánási, T., & Csermely, P. (2009). Network strategies to understand the aging process and help age-related drug design. *Genome Medicine, 1,* 90.
16. Lineweaver, C. H., Davies P. C. W., & Ruse, M. (2013). *Complexity and the arrow of time*. Cambridge: Cambridge University Press.
17. Eddington, A. S. (1981). *The nature of the physical world*. Michigan: University of Michigan Press.
18. Price, H. (2006). The thermodynamic arrow: Puzzles and pseudo-puzzles. In: I. Bigi & M. Faessler (Eds.), *Time and Matter* (pp. 209–224). Singapore: World Scientific.
19. Layzer, D. (1975). The arrow of time. *Scientific American, 233*(6), 56–69.
20. Lodge, R. C. (1956). *The philosophy of Plato*. Routledge and Kegan Paul Ltd.
21. Seneca, L. A. (2015). *Tutte le opere*. Rome: Newton Compton Ed.
22. Craig, W. L. (1980). *The cosmological argument from Plato to Leibniz*. London: Macmillan Press.
23. Feldman, F. (2000). The termination thesis. *Midwest Studies in Philosophy,* 24, 98–115.
24. Luper, S. (Ed.). (2014). *The Cambridge companion to life and death*. Cambridge: Cambridge University Press.

Chapter 4
Constant Decline

The realistic structural model (3.4), which depicts S when it is attacked by *continuous side effects,* makes clearer the principle of superposition and also the *principle of progressive deterioration.* However this principle has been expressed by words and turns out to be inappropriate for the calculus: How to qualify the progressive deterioration?

4.1 Two Introductory Remarks

Remark 1 We have already commented on the specialized parameters used to gauge the decay of artificial and natural systems, and accurate measurements provide evidence about elements which undergo regular decadence processes with time for example oxidation [1], vibration [2], corrosion [3], friction [4], overheating, overloading, and others. The concept that the component of a device wastes according to a linear trend emerges even in theoretical calculations, for example see [5, 6].

Example Railway materials—wheels and railway tracks—chafe due to trains' traveling. Donzella with others measures how the wear rate of wheels slopes down with regularity depending on the number of cycles [7]. The average wear rate is 2.74×10^{-4} μm/cicle (Fig. 4.1).

Processes appear somewhat more complicated in the biological domain as cells die and new-born cells replace the dead ones [8]. Cells are replaced over periods ranging from a few weeks to several years; however new cells do not keep the same level of vitality. For instance, biologists observe a decrease in the number of proliferative cells and a decrease in the rate of cell division over time. Cell replacement undergoes a progressive decadence with time due to collateral effects such as accumulation of insoluble particles in the tissues and a decrease in responsiveness to feedback signals.

P. Rocchi, *Reliability is a New Science*, DOI 10.1007/978-3-319-57472-1_4

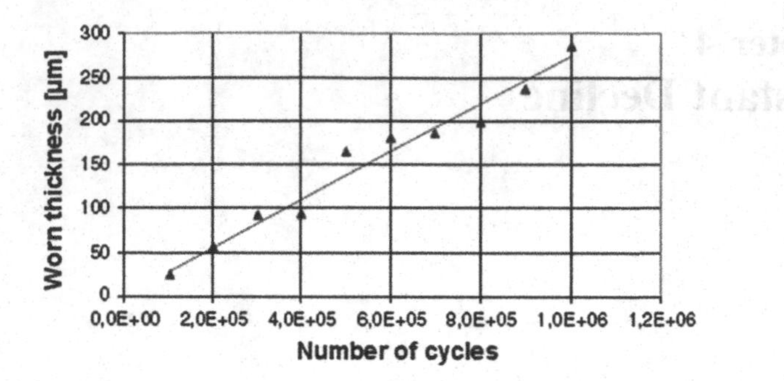

Fig. 4.1 Wear rate of railway wheels. Reproduced from [7] with permission from Author G. Donzella

The insufficient replacement of elements occurs in social systems too. It usually occurs that the members of a society—e.g. a political, economic or professional society—die or anyway leave the group and new members take their places. Sometimes new members are unable to keep alive the society. As new born cells make feebler tissues, so new born individuals make a society even more inefficient, causing it inevitably to die. When new generations have grown up with more comfortable and luxurious life standards, they become idle and dissipated. This idea was central to the *degeneration theory* with respect to which in the nineteenth century there was a certain consensus in the social and political sciences; it explained the end of the Roman empire and other political entities [9].

Examples The conviction that components waste regularly with time is rather popular in biological literature. Experimentalists bring evidence that the degeneration of an organism—after a certain age—follows a linear trend over time due to the contribution of microscopic and macroscopic components.

Ackermann and others [10] conduct tests on the reproductive function of Caulobacter crescentus which show linear slopes (Fig. 4.2 left).

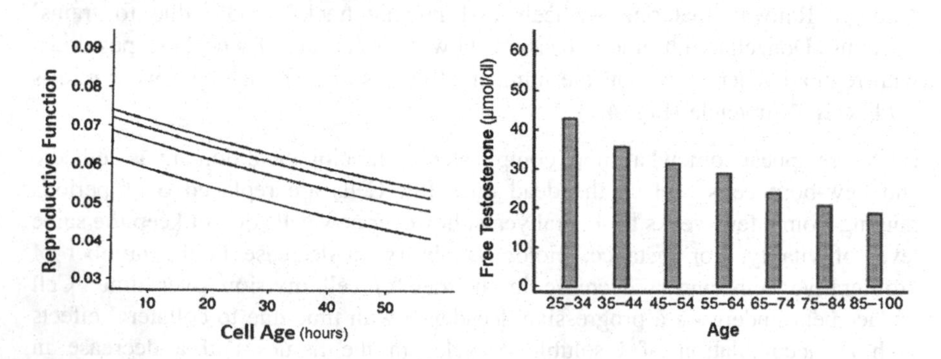

Fig. 4.2 (*Left*) Smoothed rate of cell division from four bacteria cohorts. Elaborated from [10] with permission from Wiley. (*Right*) Constant decrease of testosterone with age. Modified from [11] with permission from Springer

The reproductive functions decrease at the level of cells as time goes by and even at the level of organs. Empirical enquiries demonstrate how the testosterone concentration slopes down rather regularly at the rate of 1–2% per year (Fig. 4.2 right).

Remark 2 Whatsoever curve begins to resemble a straight line at infinitesimally close observation. A function can be approximated with a line in a finite range of the domain. 'Local linearity' can serve as a very easily computed and conceptually simple approximation of any function. The present theory has the scope of calculating $\lambda(t)$ in a limited interval of time and conforms to the 'local linearity' criterion.

In conclusion, remarks 1 and 2 suggest that a linear expression could reasonably qualify the *principle of progressive deterioration* (3.5).

Linearity Assumption: *When the sub-state A_{fg} is working, the reliability entropy H_{fg} decreases at a constant rate over time*

$$\frac{\Delta H_{fg}}{\Delta t} = \frac{[H_{fg}(t) - H_{fg}(t_0)]}{(t - t_0)} = -c_g, \qquad \Delta t \neq 0, c_g > 0. \tag{4.1}$$

For the sake of simplicity let H_{fg} $(t_0) = 0$ and $t_0 = 0$ and we get

$$H_{fg} = H_{fg}(t) = -c_g t. \tag{4.2}$$

4.2 Comments on the Linearity Assumption

Systematic wear and tear is the dominant root-cause of systems' ceasing of function. Several researchers share this stance and consider degeneration to be the essential reason of the system's undependability.

4.2.1 In the sixties, a new methodology began to estimate the unreliability of a system evaluating its performance degradation, rather than its to-failure data. Advances in sensor technologies have favored the use of probes for continuously monitoring critical elements of systems in engineering and medicine, and in turn have aided the so-called *degradation reliability methods* (DRM) which predict a system' residual life from critical performance measurements. Several theorists argue that degradation-based techniques are superior alternatives to inquiries on break-down [12]; DRM can exploit a plethora of degradation-related data that is attainable through advanced sensing technologies.

The advantages of using sensor data to estimate the future health of a component appear evident. Failure-time data may be very limited when systems are highly reliable; when it is prohibitively expensive to run systems to failure. The collection of sensor data turns out to be a faster and more agile implementation than gathering failure data which is often significantly time consuming.

4.2.2 Early ideas on DRM were given by Gertsbakh and Kordonsky [13]. The literature in the field, while relatively immature as compared to that of failure-based reliability, has grown at a rapid pace.

Sensors handle specific physical quantities, and the specific principles underlying the deterioration of each kind of system influenced the theorists in the beginning. Translating sensor data into general statistical distributions for reliability estimation remained a significant challenge. There exists a critical need for novel techniques that can map sensor data to degradation indices. An earlier example was the *proportional hazard model* (PHM) developed by Cox in 1972. Lu and Meeker [14], and Meeker with others [15] set up general statistical models to estimate the time-to-failure distribution from degradation measures.

Kiessler and others [16] investigate the limiting average availability of a system whose time-varying wear rates are governed by a continuous-time Markov model. Kharoufeh and others [17] extend the previous scheme by including damage-inducing shocks. Other stochastic failure models also attempt to capture the impact of a randomly varying environment. An excellent overview of those models, both univariate and multivariate, can be found in [18].

4.2.3 Some authors describe the progressive decadence of the generic element A_{fg} by means of the *degradation function* $D_g(t)$ [19]. This function defines an average curve interpreted as damages accumulated on the decaying unit. There are many forms of $D_g(t)$ and we confine ourselves to the most straightforward degradation function

$$D_g(t) = \xi_g t, \qquad \xi_g > 0. \tag{4.3}$$

where ξ_g is a constant depending on the component A_{fg} of the system *S*. From assumption (4.2) we have

$$t = -\frac{H_{fg}(t)}{c_g}. \tag{4.4}$$

Equations (4.3) and (4.4) prove that the degradation function matches with the reliability entropy up to a constant

$$D_g(t) = -\left(\frac{\xi_g}{c_g}\right) H_{fg}(t) = -\rho H_{fg}(t), \qquad \rho > 0. \tag{4.5}$$

This states that lowering performances of A_{fg} in (4.2) correspond to the contemporary growing degradation $D_g(t)$ of that component.

4.2.4 DRM and the present frame are strongly involved in assessing the progressive decay of physical systems, but the inquiries pursue different purposes.

Degradation testing does not need to witness any ‘hard failure’, and DRMs reduce the amount of tests, and simplify the predictions of the systems’ behavior. *Accelerated life testing* (ALT) also shares these practical purposes and some DRM can be combined with ALT. In short, DRMs pursue serviceable interests; instead

Table 4.1 The extremes of the Boltzmann entropy, the reliability entropy and the free energy

High capability for working	Low capability for working
$H_A(W)$ minimum for TS	$H_A(W)$ maximum for TS
$H(P_f)$ maximum for S	$H(P_f)$ minimum for S
F_S maximum for TS	F_S minimum for TS

the present theory is an attempt to explain broad phenomena and to establish general laws in the reliability science. It may be said that DRM is to the present logical frame, as Bayesianism is to subjectivism. As Bayesian studies demonstrate the reasonableness of the subjective interpretation of probability, which seemed arbitrary in the past; so the mounting number of DRM works shows how most failures arise from a degradation mechanism at work, and how the present approach has a solid inductive base.

4.2.5 We started with the idea that there is some connection between thermodynamics and the reliability domain. The *linearity assumption* enables us to cross these fields and compare three functions that qualify the capability of working.

The Boltzmann entropy indicates the amount of degenerated energy of A and in a way $H_A(W)$ qualifies the incapability of working of the thermodynamic state A. The physical meanings of the Boltzmann entropy and of the entropy $H(P_f)$ appear rather close, but they follow opposite trends in the sense summed up in Table 4.1.

The *free energy* (or *Helmholtz energy*) is defined as the internal energy U_S of the thermodynamic system S minus the amount of energy that cannot be used to perform that work. The unusable energy is given by the thermodynamic entropy H_S, multiplied by the temperature T of the surroundings of S

$$F_S = U_S - (H_S \cdot T). \tag{4.6}$$

The free energy F_S qualifies the amount of work that a thermodynamic system can perform. This parameter decreases in the same direction as $H(P_f)$.

4.3 Hazard Rate Caused by Constant Decline

Let us calculate the consequence from the linearity hypothesis.

Theorem 4.1 *Suppose (4.2) true for every sub-state A_{f1}, A_{f2},....A_{fn}, then the probability of good functioning until the first failure follows the exponential law with constant hazard rate*

$$P_f = P_f(t) = \exp(-ct), \qquad c > 0. \tag{4.7}$$

Proof The entropies of the sub-states are summable due to (2.16). Using (4.2) we obtain

$$H_f(t) = \sum_{g=1}^{n} H_{fg} = -\sum_{g=1}^{n} c_g t = -ct, \quad c > 0. \tag{4.8}$$

The definition of the Boltzmann-like entropy (2.26) leads to

$$H_f(t) = \ln\left(P_f\right) = -ct.$$

And (4.7) is proved.

Lemma 4.1 *The definition of hazard function* (1.6) *and* (4.7) *yield*

$$\lambda(t) = -\frac{P'(t)}{P(t)} = -\frac{(-c)\exp(-ct)}{\exp(-ct)} = c \quad c > 0. \tag{4.9}$$

When the performances of the sub-states decrease linearly with time, the hazard rate is flat.

Examples There is abundant empirical inquiries that support Theorem 4.1. Let us see three classes of devices that work all day long and the degeneration of their components conforms to (4.2) with a good degree of approximation.

Power transformers are machines used for adjusting the electric voltages to a suitable level on each segment of the power transmission framework (Fig. 4.3). Electrical current provides the regular input of the transformers; but electrical overfeeding occurs from time to time. The core, the windings, the cooling system and other components of a transformer perform regular work for years. Schiffman [20] conducts annual observations on the failures of a group of electrical power transformers from 1968 to 1982; and concludes that "the specification of a constant hazard rate is adequate".

Accurate reliability testing of electro-mechanical microengines furnished the data regarding the first period of the microengine's lifetime [21]. After the initial

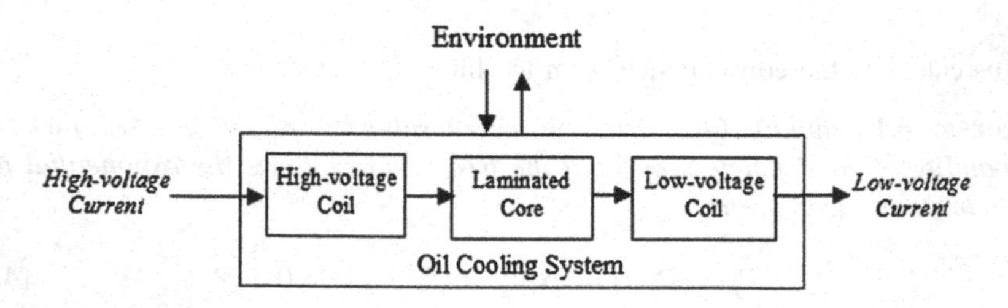

Fig. 4.3 Electric power transformer

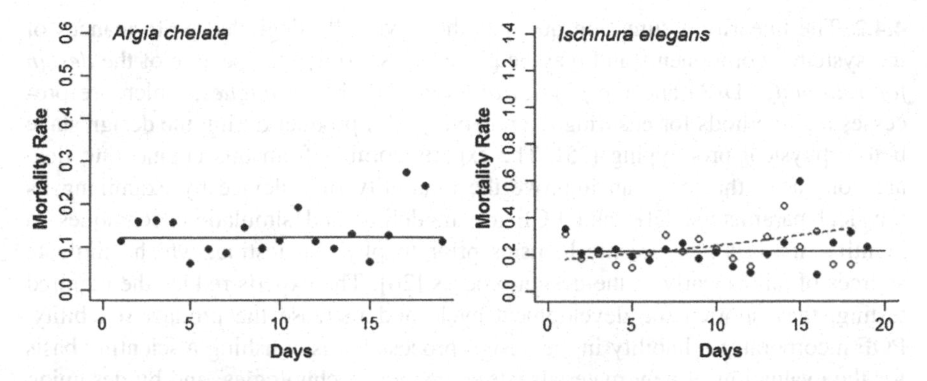

Fig. 4.4 (*Left*) Mortality rates of Argia chelata and (*Right*) Ischunura elegans male (*black bullets*) and female (*white bullet*). Reproduced from [24] with permission from Wiley

rapid descending trend of the hazard rate (infant mortality), the curve becomes nearly constant.

Design of wind turbines has rapidly evolved through time. These devices are required to operate for nearly 20 years continuously. Echavarria, Hahn and others [22] make an empirical research on the reliability of wind turbines during fifteen operational years and bring evidence that the components' failure remain rather constant.

Examples Some species of animals including hydras, insects, birds and rodents have constant mortality rates [23]. Mortality after the first year among birds is well substantiated for passerines [24]. Figure 4.4 plots the mortality rate of two species of damselflies. One can suppose that the small size of these animals minimizes the intricate deterioration mechanisms (which we shall examine in the next chapters) and thus mortality follows a rather linear trend.

4.4 Comments on Theorem 4.1 and Lemma 4.1

We have assumed that a system's components progressively degenerate and as a final consequence the hazard rate is constant

$$\textit{Linearity Assumption} \Rightarrow \boxed{\begin{array}{l} P_f(t) \textit{is exponential function;} \\ \lambda(t) \textit{is constant} \end{array}}$$

4.4.1 The present frame demonstrates something that matches with the common sense. We have proved that a regular cause—the linearity assumption—results in a regular consequence which is the constant hazard rate.

4.4.2 The linearity assumption qualifies the physically degraded performances of the system's components and may be placed close to the perspective of the *design for reliability* (DfR) and the *physics of failure* (POF) *techniques*, which are processes and methods for ensuring the reliability of a product during the design stage before physical prototyping [25]. The experts coming from this engineering field are convinced that one can improve the reliability of a device by examining its physical parameters. DfR and POF use modeling and simulation techniques to identify first-order failure mechanisms prior to physical testing, which eliminate sources of failure early in the design process [26]. The experts reduce the required testing, they shorten the development cycle and increase the product reliability. POF incorporates reliability in the design process by establishing a scientific basis for the evaluation of new materials, structures and technologies, and by designing screens and safety factors. The POF method aims to prevent failures through robust design and manufacturing practices but it cannot dream of eliminating failures. A project is based on key trade-offs between costs and technical specifications, emerging and traditional technology etc. The constant $\lambda(t)$ leads to a good compromise. The desired life span progressively worsens without dramatic interruptions.

References

1. Denis, A., & Garcia, E. A. (1988). Model to simulate parabolic followed by linear oxidation kinetics. *Oxidation of Metals, 29*(1), 153–167.
2. Lin, S. C., & Hsiao, K. M. (2001). Vibration analysis of a rotating Timoshenko beam. *Journal of Sound and Vibration, 240*(2), 303–322.
3. Barnartt, S. (1969). Linear corrosion kinetics. *Corrosion Science, 9*(3), 145–156.
4. Yanfeng, T., Huijuan, F., & Jian, Z. (2012). The life prediction model of vehicles engine based on cylinder wear law. In *2nd International Conference on Electronic & Mechanical Engineering and Information Technology* (pp. 2275–2278).
5. Ahmad, M., & Sheikh, A. K. (1984). Bernstein reliability model: Derivation and estimation of parameters. *Reliability Engineering, 8*(3), 131–148.
6. Penga, H., Fenga, Q., & Coitb, D. W. (2010). Reliability and maintenance modeling for systems subject to multiple dependent competing failure processes. *IIE Transactions, 43*(1), 12–22.
7. Donzella, G., Mazzù, A., & Petrogalli, C. (2007). Un modello fenomenologico per la previsione della competizione tra usura e RFC nel contatto ciclico tra ruota e rotaia. *Atti del Congresso IGF*19, 169–176.
8. Mizushima, N., & Komatsu, M. (2011). Autophagy: Renovation of cells and tissues. *Cell, 147*(4), 728–741.
9. Herman, A. (1997). *The idea of decline in western history*. N.Y: The Free Press.
10. Ackermann, M., Chao, L., Bergstrom, C. T., & Doebeli, M. (2007). On the evolutionary origin of aging. *Aging Cell, 6*(2), 235–244.
11. Kaufman, J. M., & Vermeulen, A. (1998). Androgens in male senescence. In E. Nieschlag & H. M. Behre (Eds.), *Testosterone: Action, deficiency, substitution* (pp. 437–471). Berlin: Springer.

12. Chiao, C. H., & Hamada, M. (2001). Experiments with degradation data for improving reliability and for achieving robust reliability. *Quality and Reliability Engineering International, 17,* 333–344.
13. Gertsbakh, I. B., & Kordonsky, K. B. (1969). *Models of failure.* Translated from the Russian version. Berlin: Springer-Verlag.
14. Lu, C. J., & Meeker, W. Q. (1993). Using degradation measures to estimate a time-to-failure distribution. *Technometrics, 35,* 161–174.
15. Meeker, W. Q., Escobar, L. A., & Lu, C. J. (1998). Accelerated degradation tests: Modeling and analysis. *Technometrics, 40,* 89–99.
16. Kiessler, P., Klutke, G., & Yang, Y. (2002). Availability of periodically inspected systems subject to Markovian degradation. *Journal of Applied Probability, 39,* 700–711.
17. Kharoufeh, J. P., Finkelstein, D., & Mixon, D. (2006). Availability of periodically inspected systems with Markovian wear and shocks. *Journal of Applied Probability, 43,* 303–317.
18. Singpurwalla, N. D. (1995). Survival in dynamic environments. *Statistical Science, 10,* 86–103.
19. Freitas, M. A., dos Santos, T. R., Pires, M. C., & Colosimo, E. A. (2010). A closer look at degradation models: Classical and Bayesian approaches. In M. S. Nikulin, N. Limnios, N. Balakrishnan, W. Kahle, & C. Huber-Carol (Eds.), *Advances in degradation modeling: Applications to reliability, survival analysis, and finance* (pp. 157–180).
20. Schiffman, D. A. (1986). The score statistic in constancy testing for a discrete hazard rate. *IEEE Transactions on Reliability, R-35*(5), 590–594.
21. Tanner, D. M., Smith, N. F., Irwin, L. W., Eaton, W. P., Helgesen, K. S., Clement, J. J., et al. (2000). *MEMS reliability: Infrastructure, test structures, experiments and failure modes* (Sandia Report, SAND2000-0091).
22. Echavarria, E., Hahn, B., van Bussel, G. J., & Tomiyama, T. (2008). Reliability of wind turbine technology through time. *Journal of Solar Energy Engineering, 130*(3), 1–10.
23. Sherratt, T. N., Hassall, C., Laird, R. A., Thompson, D. J., Cordero-Rivera, A. (2011). A comparative analysis of senescence in adult damselflies and dragonflies (Odonata). *Journal of Evolutionary Biology, 24,* 810–822.
24. Lack, D. (1954). *The natural regulation of animal numbers.* Wotton-under-Edge: Oxford Clarendon Press.
25. Chatterjee K., Modarres M., & Bernstein J. (2012). Fifty years of physics of failure. *Journal of the Reliability Analysis Center, 20*(1).
26. Matic, Z., & Sruk, V. (2008). The physics-of-failure approach in reliability engineering. In *Proceedings of the International Conference on Information Technology Interfaces* (pp. 745–750).

Chapter 5
Random Factors

The present chapter means to discuss the interruptions of the system operations due to a *random factor* (RF). The first group of failure determinants consists of continuous mechanisms; RFs make the second group. The illustration of the present argument mirrors the modern literature and does not put forward original solutions in terms of mathematics. This chapter handles random factors in the traditional way and is more concise than the previous chapter for this reason.

5.1 Random Factors of Failure

We examine RFs that prompt the definitive cessation of the system functions. RFs intervene at any time during the system lifetime and their occurrence does not follow any pattern. Failures occur purely by chance and some writers use the colored expression 'acts of God'.

5.1.1 Random determinants may have an intrinsic origin or an extrinsic origin, or both. Manufacturing defects, specification mistakes and implementation errors are examples of anomalies inherent to systems which can cause unforeseen interruptions [1]. Examples of external agents are involuntary human errors or alternatively, intentional acts such as sabotage, destruction etc. Other external dangerous elements can come from the environment at random [2].

A casual element of failure basically works in two alternative ways. It results in:

- *An immediate upshot*, e.g. a stone hits the windows glass and directly smashes it; e.g. a bacterial infection causes the death of a patient.
- *A delayed upshot,* in accordance with the *latent failure theory* by Reason discussed in 3.1.3.

P. Rocchi, *Reliability is a New Science*, DOI 10.1007/978-3-319-57472-1_5

5.1.2 The RFs under examination come in a very broad variety of forms, but they share a concise property. They induce a sudden and unforeseen accident which can be described in the following manner:

$$\textit{Principle of Random Accidents: Once in a while the system fails due to one or more random factors which lead to S to ceasing to function.} \tag{5.1}$$

5.2 Comments on the Principle of Random Accidents

Let us place the systematic and the random factors, discussed in Chap. 4 and this chapter, side by side.

The *principle of progressive deterioration* emerges as an ineluctable truth in engineering and biology. The irrepressible everlasting actions of CF operate in such a manner as could not be otherwise. The same cannot be said for the principle (5.1) since several appropriate policies are often able to safe guard appliances and living beings from negative influence with success. One can protect the system from external accidents, or even can design and maintain it using appropriate policies. Very many long-term systems do not run into sudden accidents.

Therefore the *principle of progressive deterioration* comes into view as a mechanism that affects the entire universe, whereas the *principle of random accidents* does not share the same status.

5.3 Behavior of Accidents

For the sake of simplicity, authors assume that fortuitous failures occur by chance during a specific period of lifetime [3], and they split this period into short time intervals under the hypothesis that:

(a) The accident can occur only once per time interval,
(b) The accident occurs with the same probability in each time interval,
(c) The occurrence of the accidents takes place independently from each other in the time intervals.

Writers approximate the process—described in (a), (b) and (c)—to a binomial model since RF results in a small number of failures per unit of time, while the test time is relatively large.

Binomial Assumption: *Accidents comply with the binomial distribution that has the following mass function*

$$P(X=k)=\frac{n!}{k!(n-k)!}p^k(1-p)^{n-k}, \qquad k=0,1,2,\ldots n. \tag{5.2}$$

where n is the number of trials, p is the probability of accident in a single trial and k the number of accidents. Essentially, any time interval is an event that can be either a success or a failure. When the sample size is unknown, authors adopt the Poisson distribution which can be derived from (5.2).

Theorem 5.1 *Suppose (5.2) true and*

$$ft=np, \qquad f>0. \tag{5.3}$$

where t is the given time interval and f is a positive constant. If $n\rightarrow\infty$, then the probability mass function becomes

$$P(X=k)=\frac{(ft)^k}{k!}\exp\left(-ft\right). \tag{5.4}$$

Proof From (5.3) we get

$$p=ft/n.$$

We substitute this expression for p into the binomial distribution (5.2)

$$P(X=k)=\frac{n!}{k!(n-k)!}\left(\frac{ft}{n}\right)^k\left(1-\frac{ft}{n}\right)^{n-k}. \tag{5.5}$$

We pull out the constants f^k and $1/k!$ and calculate the following limit

$$\frac{(ft)^k}{k!}\lim_{n\rightarrow\infty}\frac{n!}{(n-k)!}\left(\frac{1}{n^k}\right)\left(1-\frac{ft}{n}\right)^n\left(1-\frac{ft}{n}\right)^{-k}. \tag{5.6}$$

This limit can be subdivided into three parts. The first limit consists of k fractions that tend to the unit when n approaches infinity

$$\lim_{n\rightarrow\infty}\frac{n!}{(n-k)!}\left(\frac{1}{n^k}\right)=\lim_{n\rightarrow\infty}\left(\frac{n}{n}\right)\left(\frac{n-1}{n}\right)\left(\frac{n-2}{n}\right)\cdots\left(\frac{n-k+1}{n}\right)=1 \tag{5.7}$$

The limit of the term in the middle of (5.6) is the following because of a property of the Neper number

$$\lim_{n\rightarrow\infty}\left(1-\frac{ft}{n}\right)^n=\lim_{x\rightarrow\infty}\left(1-\frac{1}{x}\right)^{x(-ft)}=\exp\left(-ft\right). \tag{5.8}$$

The last limit is one as n approaches infinity

$$\lim_{n\to\infty}\left(1-\frac{ft}{n}\right)^{-k}=1. \tag{5.9}$$

We place partial results (5.7), (5.8) and (5.9) into (5.6) and the theorem is proved

$$\begin{aligned}
&\frac{(ft)^k}{k!}\lim_{n\to\infty}\frac{n!}{(n-k)!}\left(\frac{1}{n^k}\right)\left(1-\frac{ft}{n}\right)^n\left(1-\frac{ft}{n}\right)^{-k}\\
&=\frac{(ft)^k}{k!}(1)[\exp(-ft)](1)\\
&=\frac{(ft)^k}{k!}[\exp(-ft)].
\end{aligned}$$

The inter-arrival times of accidents are irrespective of one another and constitute a homogeneous Poisson process [4].

5.4 Hazard Rate Caused by Random Accidents

The outcome just obtained leads to the next theorem.

Theorem 5.2 *Suppose* (5.4) *true, the probability of good functioning until the first failure is the following exponential.*

$$P_f = P_f(t) = \exp(-ft), \qquad f > 0 \tag{5.10}$$

Proof The device stops at the first accident, thus one poses $k = 0$ in (5.4) and gets the probability of no failure in the interval (0, t).

$$P_f = P_f(t) = P(k,t) = \frac{1}{1}\exp(-ft) = \exp(-ft), \qquad f > 0. \tag{5.11}$$

The probability of good functioning P_f follows the exponential law.

Lemma 5.1 *Using definition* (1.6) *we get the hazard rate is constant*

$$\lambda(t) = f, \qquad f > 0. \tag{5.12}$$

Example The life span of one hundred terminals used in a software house is exponentially distributed. Managers register about 5 failures per year thus the probability of good functioning after 5 years is

$$P_f = \exp(-0.05\cdot 5) = 0.778801$$

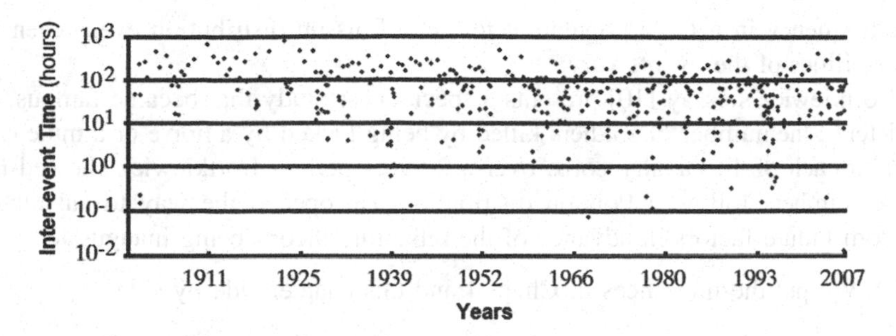

Fig. 5.1 Inter-event times of global tsunami catalog. Reproduced from [7] with permission from Harvard University Press, Copyright © 2009

Table 5.1 Cervical cancer age-rate per 1000 inhabitants. Reproduced from [8] with permission from Springer

Denmark	16
England and Wales	12
USA (White)	7
USA (Black)	12

Example For decades the Poisson distribution has been employed to study rainfalls, thunderstorms and other climatic accidents [5, 6]. More recently the Poisson model has supported extensive inquiries on predicting earthquake and tsunami (Fig. 5.1).

Example In epidemiology some researchers assume that the hazard function is constant over a certain period of time. For instance, cancer incidence data are given in five-year age bands. The likelihood of such exponential models can be written simply and provide good support for statistical analysis (Table 5.1).

Example Sudden events can subvert a political system. Critical events may take the form of scandals, assassinations, economic calamities, social or cultural upheavals, international incidents, voter referenda, strikes, interparty disputes etc. These heterogeneous situations have an exogenous and endogenous nature and can be treated by mathematical models [9].

5.5 Comments on Theorem 5.2 and Lemma 5.1

We mention an historical case.

5.5.1 Ladislaus von Bortkiewicz (1868–1931) lived most of his life in Germany and taught at Strasbourg University. He was involved in discussing the results of performing the same experiment a small number of times. He noted that events with low frequency in a large population follow a Poisson distribution even when the probabilities of the events vary.

Bortkiewicz's essay [10] presents a special case study that became famous. He registered the number of soldiers killed by being kicked by a horse or a mule each year in each of 14 cavalry corps over a 20-year period. Bortkiewicz showed that those numbers follow a Poisson distribution. He opened the way to calculating random failure factors in advance of the reliability theory being inaugurated.

5.5.2 We put the inferences of Chap. 4 and this chapter side by side.

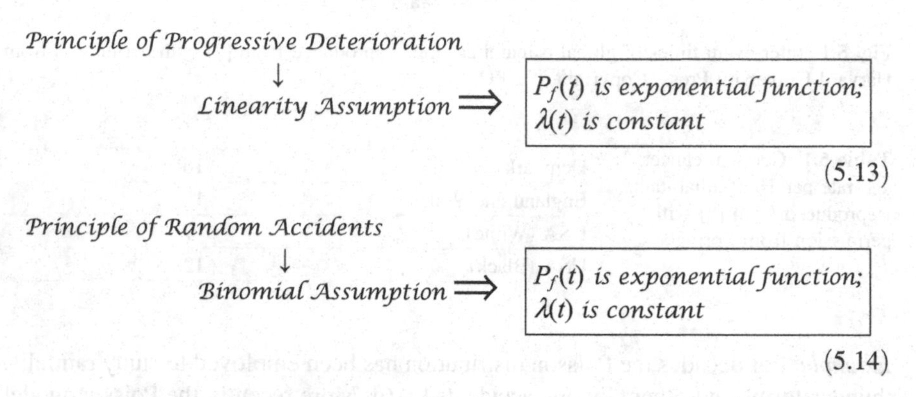

These logical implications constitute an extraordinarily fine result since an identical conclusion is derived from far different premises. Continuous and random factors which can have dissimilar origins and forms are governed by a single rule: the exponential law.

5.5.3 The principle of superposition holds that an assortment of factors contributes to a single fault. Authors who explore the so-called *competing causes* of failure, subdivide them into *wearout* and *overstress* mechanisms [11]. The first determines the progressive damage accumulation; the second produces instantaneous catastrophic failure whenever they are severe enough to exceed the strength of the material. Wearout and overstress mechanisms mirror the subdivision (5.13) and (5.14).

5.5.4 We can summarize the virtues of the results (5.13) and (5.14) which offer the following significant advantages:

I **Simplicity**: The constant hazard rate is the most straightforward model and experts can address several reliability issues with ease.

II **Indifference**: A sudden accident occurs in parallel to the long-term deterioration and does not modify the type of $\lambda(t)$ that is typically constant.

III **Uniformity**: Systematic and random factors may be intrinsic, extrinsic or both. When they overlap this variety of determinants does not compound statistical studies that employ only the exponential model. The enquiries do not need to disentangle the failure causes as they obey the same rule.

IV **Coverage**: Wear out and random attacks embody the most common failures and can occur during the entire system lifetime, thus the exponential distribution covers a wide spectrum of situations in abstract and in applications.

5.5.5 In summary, the following plain mathematical model dominates the reliability domain

$$\lambda(t) = \text{const.}$$

The present comprehensive frame justifies the weight of the exponential law from the theoretical viewpoint while the remarks from I to IV detail the success that the exponential model has obtained in the community of statisticians long since [12]. Engineers, economists, doctors and others are inclined to use the constant hazard rate model in the quest for simplicity. Linearity can serve as a hazard rate that is somewhat easy to understand and carry out; it exhibits good qualities in terms of exhaustiveness and applicability.

5.5.6 Using (5.10) and supposing t_0 to be any age, the conditional probability is

$$\begin{aligned} P_f(t/t_0) &= \frac{P(t_0+t)}{P(t_0)} \\ &= \frac{\exp\left[-f(t_0+t)\right]}{\exp\left[-f(t_0)\right]} = \exp\left(-ft\right). \end{aligned}$$

The distribution of the system's remaining survival time, given that it has survived until time t_0, does not depend on t_0. This is usually called *memoryless property.*

5.6 Shock Theories

The present chapter is somewhat close to the *shock models.* RFs result in sudden failures or even strikes against S, the current literature marks these events with the term '*shocks*' or even '*shock loads*'. Authors adopt this terminology for man-made systems and for biological systems as well though the medical definition of 'shock' has a very different meaning. Medically, 'shock' is defined as a condition where the body does not receive enough oxygen and nutrients.

5.6.1 *Shock theories* describe catastrophic failures using the models of stochastic processes. Experts search to relate the interval between two consecutive shocks, the damage size caused by individual random shocks and the system failure function [13]. Traditionally, theories are of two kinds:

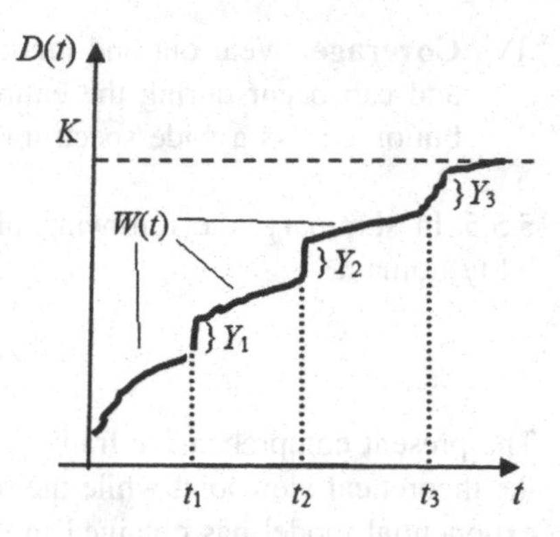

Fig. 5.2 The total degradation process under random shocks

- *Extreme shock models* investigate the accidents caused by a shock that is larger than some critical level.
- *Cumulative shock models* relate the failure of S to the cumulative effect of a large number of shocks.

Further extensions including the *δ-shock model, mixed shock model* and *run shock model* have been developed in recent years.

5.6.2 A single extreme shock load is capable of destroying S. Alternatively, it may happen that a blow is not enough but a certain number of blows are necessary to bring the system to an end. The impact of this disturbance accumulates in accordance with the principle of superposition [14].

Suppose that $N(t)$ is the number of shocks that arrive according to a homogeneous Poisson process, and the shock damage size, denoted as Y_j, qualifies the instant increase in degradation [15]. The cumulative damage size of the degeneration process due to random shocks until t is given as

$$S(t) = \sum_{j=1}^{N(t)} Y_j, \qquad j = 1, 2, \ldots \infty.$$

The total degradation function $D(t)$, which does not exceed a certain threshold K (Fig. 5.2), includes ageing decay $W(t)$ and the cumulative degeneration $S(t)$

$$\begin{aligned} D(t) &= W(t) + S(t) \\ &= W(t) + Y_1 + Y_2 + \cdots + Y_{N(t)}. \end{aligned}$$

The overall process makes a stepwise trend. It may be said that the linear degradation constitutes the base-line while the shocks accelerate the linear degradation process. Cumulative shock models prove to be in agreement with the progressive decay model examined in the previous chapter.

References

1. Juran, J. M. (1988). *Juran's quality control handbook* (4th ed). New York: McGraw-Hill.
2. Ebeling, C. E. (1997). *An introduction to reliability and maintainability engineering*. New York: McGraw-Hill.
3. Gut, A., & Hüsler, J. (1999). Extreme shock models. *Extremes, 2*(3), 295–307.
4. Knight, K. (1999). *Mathematical statistics*. Boca Raton, Florida, Chapmann & Hall/CRC.
5. Grant, E. L. (1938). Rainfall intensities and frequencies. In *Transactions of the american society of civil engineers* (Vol. 103, pp. 384–388).
6. Thompson, C. S. (1984). Homogeneity analysis of a rainfall series: An application of the use of a realistic rainfall model. *Journal of Climatology, 4,* 609–619.
7. Geist, E. L., Parsons T., ten Brink, U. S., & Lee, H. J. (2009). Tsunami probability. In E. N. Bernard & A. R. Robinson (Eds.), *The global coastal ocean*: *The sea tsunamis* (Vol. 15). Cambridge: Harvard University Press.
8. Sasieni, P. D. (2005). Survival analysis. In W. Ahrens & I. Pigeot (Eds.), *Handbook of Epidemiology* (pp. 394–728). Berlin: Springer.
9. Marchant-Shapiro, T. (2014). *Statistics for political analysis*: *Understanding the numbers*. Thousand Oaks: SAGE Publications.
10. Bortkiewicz, L. (1898). *Das Gesetz der kleinen Zahlen* [*The law of small numbers*]. Leipzig: B. G. Teubner.
11. Dasgupta, A., Sinha, K., & Herzberger, J. (2013). Reliability engineering for driver electronics in solid-state lighting products. In van Driel, W. D. & Fan, X. J. (Eds.), *Solid state lighting reliability: Components to systems* (pp. 243–284). Berlin: Springer.
12. Davis, D. J. (1952). An analysis of some failure data. *Journal of the American Statistical Association, 47*(258), 113–150.
13. Nakagawa, T. (2007). *Shock and damage models in reliability theory*. Berlin: Springer.
14. Gut, A. (1990). Cumulative shock models. *Advances in Applied Probability, 22*(2), 504–507.
15. Hao, H., Su, C., & Qu, Z. (2013). Reliability analysis for mechanical components subject to degradation process and random shock with wiener process. In *Proceeding of the 19th International Conference on Industrial Engineering and Engineering Management* (pp. 531–543).

Chapter 6
Accelerated Decline

The *linearity assumption* holds that the parts of *S* progressively degrade due to wear and tear, and this dominant phenomenon has multiple consequences. Any endangered part harms close components and triggers a *cascade* (or *waterfall*) *effect*. This compound failure mechanism accelerates the system deterioration towards the definitive end-point.

Example Let a new petrol engine begins to run. After the first movement the attrition, which is a typical factor of failure (CSE), comes into action and begins to erode the surfaces of the cylinder, the piston and other gear mechanisms. During the useful life the parts proceed to work while the performance of the petrol engine worsens (Fig. 6.1). The progressive wear and tear proceeds all the way down and the components begin to work in an incorrect manner. For instance, when the needle bearings wear, the crankshaft makes transversal movements in addition to the regular turn. This shortcoming further harms the extremes of the connecting rod that deviates from its regular movement; in turn the asymmetric action of the rod endangers the piston and in turn this endangers the cylinder. A cascade effect accelerates the engine degeneration and causes the engine to cease functioning.

Examples that are typical of the biological domain will follow in the next pages.

Frequently one cannot examine the microscopic changes of the system components, it is easier to measure the macro effects resulting from those changes. For example, mechanical wearing results in the vibration of the parts; active electrical circuits overheat. Authors have unified the intrinsic and extrinsic determinants that aggravate the status of *S* and marked them as '*degradation accelerating factors*'. We confine ourselves to quoting the following empirical works [1, 2, 3].

P. Rocchi, *Reliability is a New Science*, DOI 10.1007/978-3-319-57472-1_6

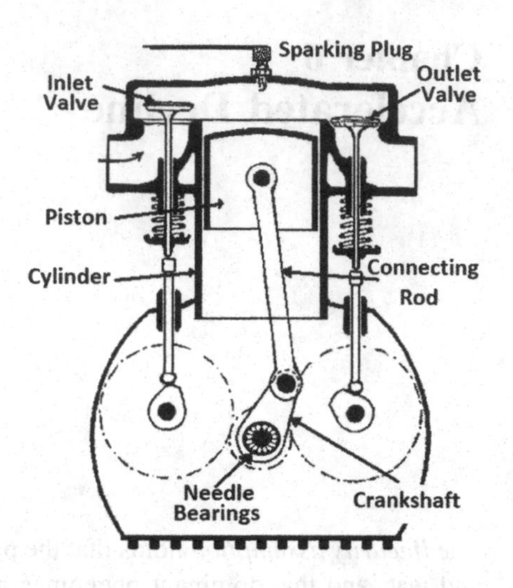

Fig. 6.1 Principal components of the petrol engine

6.1 Compound Degradation

Let us recall how n sub-states ensure the correct functions of S

$$S = A_f = (A_{f1}\, AND\, A_{f2}\, AND \ldots AND\, A_{fn}). \quad (3.1)$$

The relation *AND* entails that any part works together with the other one and therefore a degenerated part spoils a neighboring part and this in turn can affect another part and so on.

6.1.1. The progressive expansion of the cascade effect is capable of causing the definitive stop of operations. We mean to qualify this mechanism by introducing the following three hypotheses.

(a) A cascade effect results in global damaging of A_f that is to say the macro-state A_f deteriorates depending on the deterioration of the sub-states (A_{f1}, A_{f2}, ... A_{fn}). In harmony with the monotonicity axiom (2) of the Boltzmann-like entropy, we assume that the *reliability entropy of the cascade effect* H_f^* decreases in function of the entropies of the sub-states

$$H_f^* = f\left(H_{f1}, H_{f2}, \ldots, H_{fn}\right) \quad (6.1)$$

is a monotonically decreasing function of $H_{fg}\,(g = 1, 2, \ldots, n)$.

(b) A part is capable of triggering the cascade mechanism since it runs in non-perfect conditions. The generic sub-state A_{fg} has a certain degree of deterioration and also the overall functioning state A_f has a certain level of ineffectiveness. We can conclude that the cascade mechanism occurs when *the reliability entropies have reached a critically low value.* Since the maximum value of the Boltzmann-like entropy is zero, we suppose that H_{fg} is lower than zero and, for the sake of simplicity, we fix the entropic critical value is the negative unit

$$\text{If}\left|H_{fg}\right| = 1, \text{ then}\left|H_f^*\right| = 1. \tag{6.2}$$

(c) The generic sub-state A_{fg} is ineffective, and in consequence of this degenerated state it can lead S to stop functioning. This means that if only one component gives away, the overall system may give away. The definitive end of the functioning state A_f corresponds to the negative infinite entropy H_f, hence we assume that if *the entropy* H_{fg} *of the sub-state* A_{fg} *reaches the minimum, then the reliability entropy* H^*_f *reaches the minimum value*

$$\text{If } H_{fg} = -\infty, \text{ then } H_f^* = -\infty. \tag{6.3}$$

Theorem 6.1 *Suppose assumptions Eqs.* (6.1), (6.2) *and* (6.3) *are true, the reliability entropy of the system influenced by the cascade effect is the following*

$$H_f^* = -\prod_{g=1}^{n} \left|H_{fg}\right|. \tag{6.4}$$

Proof Multiplication is a repeated addition of multipliers, thus constraint Eq. (6.1) is satisfied. It is sufficient that a sole multiplier is infinite and for it the product Eq. (6.4) is infinite in accordance with assumption Eq. (6.3). We use (2.5) with assumption Eq. (6.2) and this product proves the last constraint

$$-|1 \cdot 1 \cdot 1 \cdot \ldots \cdot 1| = -1.$$

6.1.2. The entropy H_f^* qualifies the worsening mechanism achieved by n sub-states but the specific interactions amongst the parts have not been described. We are going to calculate the cascade effect and the hazard rate for two different structures.

6.1.3. The cascade effect does not stop the usual deterioration of the system components, in other words Eq. (4.2) is true when we shall calculate the next results.

6.1.4. We recall that S and A_f coincide in (3.1), and the symbols 'S' and 'A_f' are synonymous here.

6.2 Hazard Rate Caused by a Linear Cascade Effect

Let us begin with the simplest case that happens when A_f *has a linear structure*. The waterfall effect necessarily progress in a single direction.

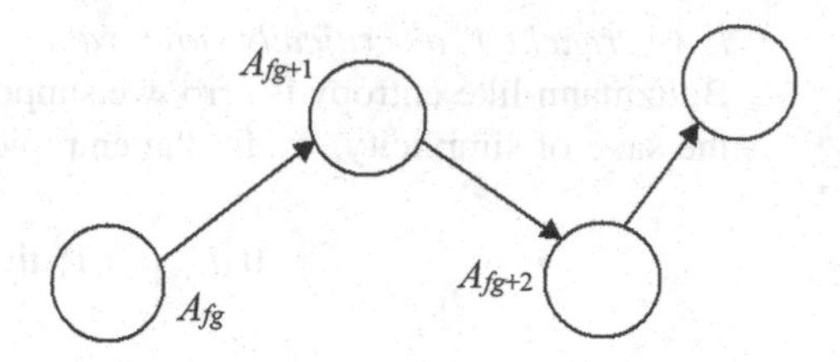

Fig. 6.2 Linear cascade effect

Linear Cascade Hypothesis: *The cascade effect develops straight in a linear structure.* (6.5)

In substance, the sub-states of S make a sequence and the spoiled sub-state $A_{fg}(g = 1, 2, .., n)$ is capable of damaging only the subsequent A_{fg+1}, this in turn harms A_{fg+2} and so forth (Fig. 6.2).

Let us calculate the hazard rate deriving from the *linear cascade hypothesis.*

Theorem 6.2 *Suppose Eqs.* (6.4) *and* (6.5) *are true, the reliability entropy* H_f^- *for the state* A_f *is the following*

$$H_f^- = -\frac{|H_{f1} \cdot H_{f2} \cdot H_{f3} \cdot H_{fn}|}{n!}. \tag{6.6}$$

Proof The result can be derived on the basis of two remarks. As first the entropy of the generic cascade effect H_f^* calculates $n!$ permutations without repetitions, whereas a linear process consists of only one permutation. Secondly, there are no special hypotheses on the cascade effect, in particular there are no special reason that unbalances the system degeneration. Concluding, we demand that the decay process is constant along the line and we divide H_f^*, defined in Eq. (6.4), by the factorial and obtain the reliability function H_f^- for the linear chain

$$H_f^- = -\frac{\prod_{g=1}^{n} |H_{fg}|}{n!} = -\frac{|H_{f1} \cdot H_{f2} \cdot H_{f3} \cdot \ldots \cdot H_{fn}|}{n!}. \tag{6.7}$$

Theorem 6.3 *Let Eqs.* (4.2) *and* (6.6) *be true, the probability of good functioning for a system impaired by the linear cascade effect is the exponential-power function*

$$P_f = P_f(t) = b\ \exp(-at^n), \quad a,\ b,\ n > 1. \tag{6.8}$$

Proof We make explicit the entropies H_{f1}, H_{f2}, H_{f3} ... in Eq. (6.6) using the *linearity assumption*

$$H_{fg} = H_{fg}(t) = -c_g t. \tag{4.2}$$

And we get

$$H_f^-(t) = -\frac{|c_1 t \cdot c_2 t \cdot c_3 t \cdot \ldots \cdot c_n t|}{n!}. \tag{6.9}$$

For the sake of simplicity we take all the constants equal to c, and we have

$$H_f^-(t) = -\frac{|c^n \cdot t^n|}{n!} = -\left(\frac{c^n}{n!}\right) t^n = -at^n. \tag{6.10}$$

where

$$\left(\frac{c^n}{n!}\right) = a.$$

By definition the entropy is the logarithm function

$$H_f^-(t) = \ln(P_f) = -at^n.$$

Thus the following holds

$$P_f = P_f(t) = b \exp(-at^n), \quad a, b, n > 1.$$

Equation (6.8) is proved where b is a suitable constant.

Lemma 6.1 *The definition of the hazard function (Eqs.* 1.6 *and* 6.8*) lead to*

$$\lambda(t) = -\frac{P'(t)}{P(t)} = -\frac{(-abt^{n-1})\exp(-at^n)}{b \exp(-at^n)} = at^{n-1}, \quad a, b, n > 1. \tag{6.11}$$

The probability of good functioning for a system affected by the linear cascade effect complies with the exponential-power law; and the hazard rate is a power of time.

6.3 Comments on Theorem 6.3 and Lemma 6.1

Engineers learn the following rule of thumb: *the simpler the design, the more reliable and less maintenance-intensive will be the final product.* Complicated appliances very likely involve a non-negligible amount of risk. Designers are inclined to arrange linear structures and the linear cascade effect is a common phenomenon in man-made systems.

6.3.1. The graphs that depict the largest work organizations in industry, agriculture, and commerce include a sequel of edges or are trees.

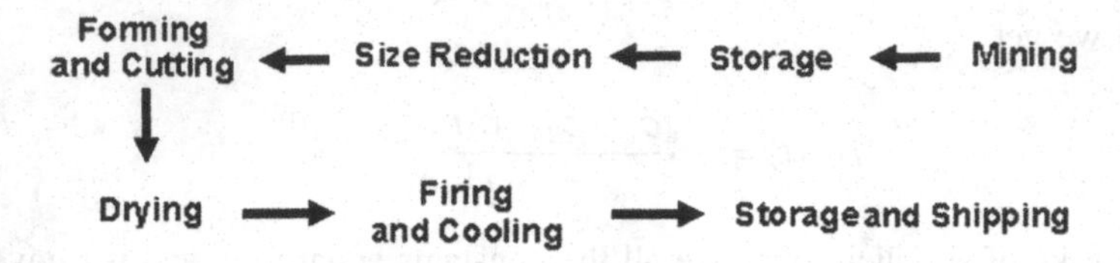

Fig. 6.3 Manufacturing process of bricks

Example The manufacturing process of silica bricks transforms raw clay into bricks through a sequence of operations (Fig. 6.3).

Sometimes experts use the *Sankey diagram* to emphasize the linear flows which substantiate the whole process of production [4]. The width of each arrow is proportional to the input/output quantity; the height of each block is directly proportional to energy absorbed or the time spent by the process or other.

Example The flow in Fig. 6.4 illustrates the production of renewable jet fuel from biomass.

6.3.2. Waloddi Weibull was a Swedish engineer who investigated the strength of materials, the fatigue and rupture in solids and bearings. In 1951 he presented a paper to the American Society of Mechanical Engineers that illustrated a statistical distribution and seven case studies. In reality, this distribution—named after Weibull—was first identified by Fréchet in 1927 and shortly thereafter applied by Rosin and Rammler to describe a grain size distribution.

The Weibull distribution includes α, β and μ, known as the *scale, shape* and *location* parameters respectively and turns out to be symmetrical to Eq. (6.8) for $\beta > 1$

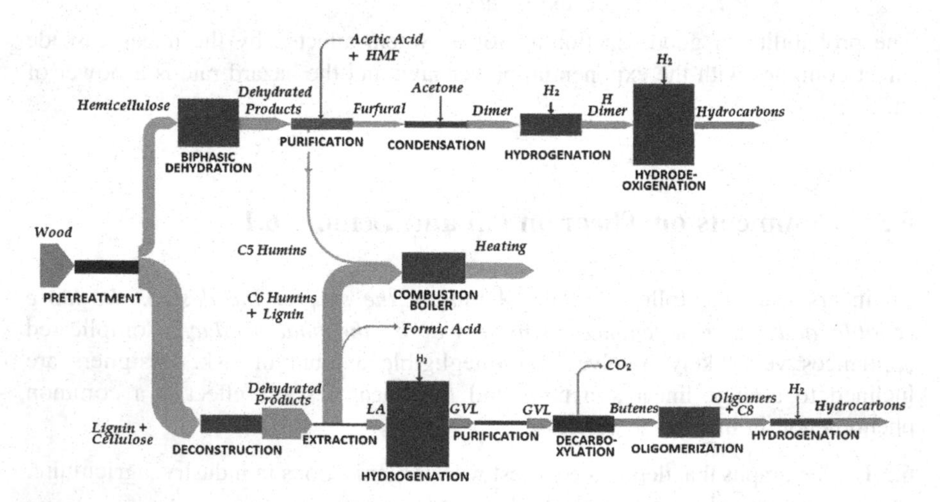

Fig. 6.4 Production of jet fuel. Adapted from [5] with permission from The Royal Society of Chemistry

$$f(t;\alpha,\beta,\mu) = \frac{\beta}{\alpha}\left(\frac{t-\mu}{\alpha}\right)^{\beta-1}\exp\left[-\left(\frac{t-\mu}{\alpha}\right)^{\beta}\right] \quad t > \mu, \alpha > 0, \beta > 1.$$

The hazard rate is a power of time and fits with Eq. (6.11)

$$\lambda(t) = \frac{\beta}{\alpha}\left(\frac{t-\mu}{\alpha}\right)^{\beta-1} \quad t > \mu, \alpha > 0, \beta > 1. \tag{6.12}$$

6.3.3. Several contemporary authors share the following opinions:

(b1) The linear disposition of operations is a typical feature of man-made systems [6, 7] in contrast to the complexity of animals and plants.
(b2) The Weibull distribution provides a good account of the distinctive reliability of appliances, equipment, devices etc., and it is so frequently used in reliability engineering that it is considered a standard mathematical tool.

6.3.4. In short, the previous pages develop the ensuing logical implication

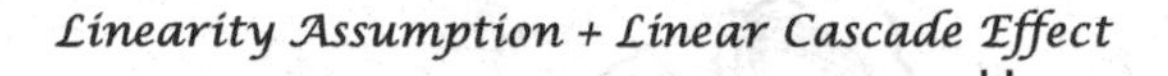

$\Downarrow$

$P_f(t)$ is exponential-power function;
$\lambda(t)$ is power function.

This implication is consistent with statements (b1) and (b2) and in addition demonstrates how (b2) depends on (b1). Theorem 6.3 proves that the reliability of *S* is an exponential-power of time when the cascade effect operates in appliances whose structure looks like a line.

6.3.5. Let us clarify Lemma 6.1 through the use of evidence furnished by the current literature.

Example Sheikh and others [8] monitored 44 pumps in an oil refinery over a period of five years. These appliances—produced by various companies—have different forms but the essential parts of the pumping systems include electrical power switches, an electrical engine, coupling components, a pump, and a set of flow control valves. These parts are operationally aligned so that they bring forth a linear hydraulic flow which complies with assumption Eq. (6.5)

Sheikh analyzes various failure modes including mechanical seal, coupling distortion, malfunction of the control valves, corrosion and cracks. The distribution of seal failures in Fig. 6.5 fits with the Weibull function when β equals 1.66.

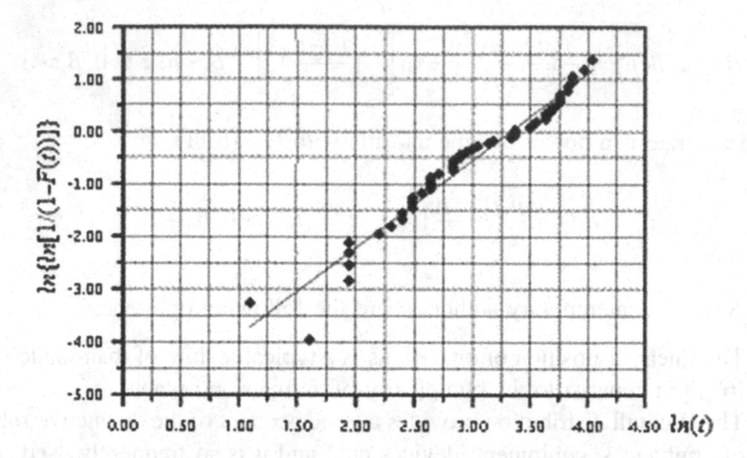

Fig. 6.5 Exponential-power distribution of seal failures. Reproduced from [8] with permission from Author A.K. Sheikh

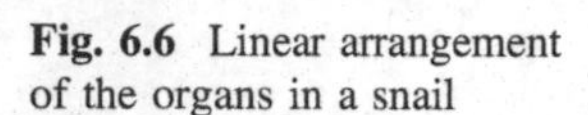
Fig. 6.6 Linear arrangement of the organs in a snail

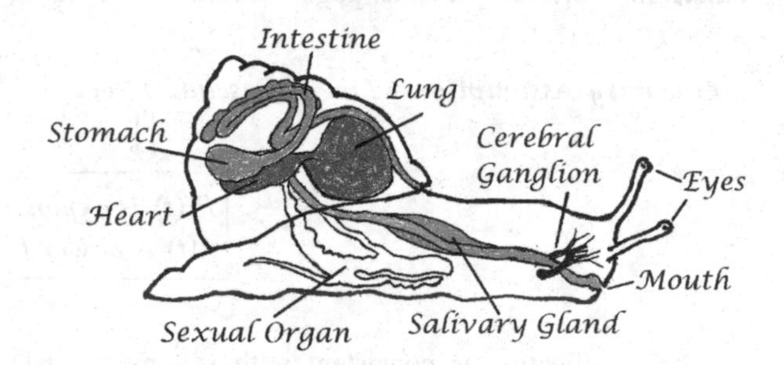

Examples Biosystems may have somewhat simple structures. The organs have a linear shape and are approximately placed one after the other. The cascade effect can but develop in line and those organisms follow the exponential-power law. The literature brings evidence that the previous conclusion is true for some species of snails [9] (Fig. 6.6) and flies [10] (Fig. 6.7). Duyck and others investigate four species of flies whose mortality matches with the Weibull distribution with the shape parameters listed in Table 6.1 [11] .

Fig. 6.7 Linear arrangement of the organs in a fly

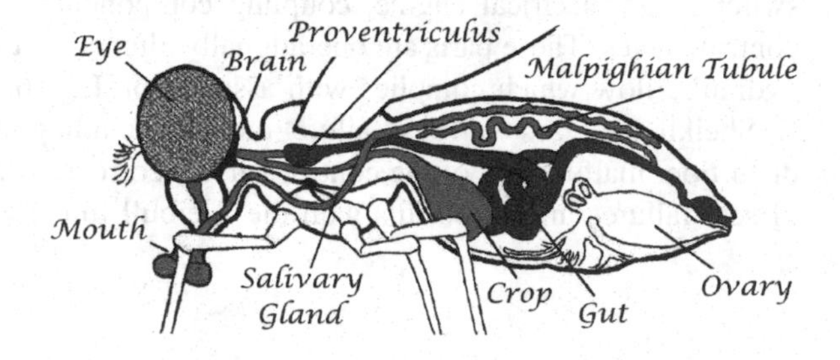

Table 6.1 Beta parameters of flies

	Ceratitis catoirii	Ceratitis capitata	Ceratitis rosa	Bactrocera zonata
Parameter β	1.366	1.485	1.422	1.265

6.3.6. When $\beta = 1$ we obtain the exponential distribution as a special case. The exponential function is nested into the Weibull, hence one could wonder:

Is Theorem 6.3 redundant?

Speaking in general, the purposes of a principle-based theory widely differ from those of statistical studies. The latter search flexible distributions that include the functions $\mathcal{B}_1, \mathcal{B}_2$, etc., whereas the former means to forecast the phenomenon $\mathcal{B}_1$, on the basis of the precise root-causes $\mathcal{A}_1$, subsequently it develops $\mathcal{A}_2 \Rightarrow \mathcal{B}_2$, $\mathcal{A}_3 \Rightarrow \mathcal{B}_3$ and so forth. More specifically, in this book Theorem 4.1 demonstrates $\lambda(t)$ when the sub-states regularly degenerate; Theorem 6.3 exemplifies the hazard rate which is affected from the previous mechanism and in addition the parts injure one another. The distinct hypotheses of Theorems 4.1 and 6.3 provide different conclusions and prevent us from merging the outcomes into a unique mathematical expression.

6.4 Hazard Rate Caused by Complex Cascade Effect

As the complexity of S increases, the potential for faults increases due to the combination of failures. In principle each component of a complicated system is able to harm not only the succeeding element but also the elements placed all around it. A single part can impair a variety of parts due to interactions moving in various directions. The complex cascade effect, an alternative to the linear cascade effect, results in the compound degeneration process.

6.4.1. We qualify the present phenomenon with the following assumption, which is mutually exclusive with assumption Eq. (6.5).

Complex Cascade Hypothesis: *The cascade effect progresses in different directions from every single sub-state.*

(6.13)

Let us look at the details of this assumption.

$$\begin{aligned}
&A_{fg}\\
&A_{fg} \rightarrow A_{fg+1}\\
&A_{fg} \rightarrow A_{fg+1} \rightarrow A_{fg+2}\\
&A_{fg} \rightarrow A_{fg+1} \rightarrow A_{fg+2} \rightarrow A_{fg+3}\\
&\dots\\
&A_{fg} \rightarrow A_{fg+1} \rightarrow A_{fg+2} \dots A_{fg+n-1}
\end{aligned} \tag{6.14}$$

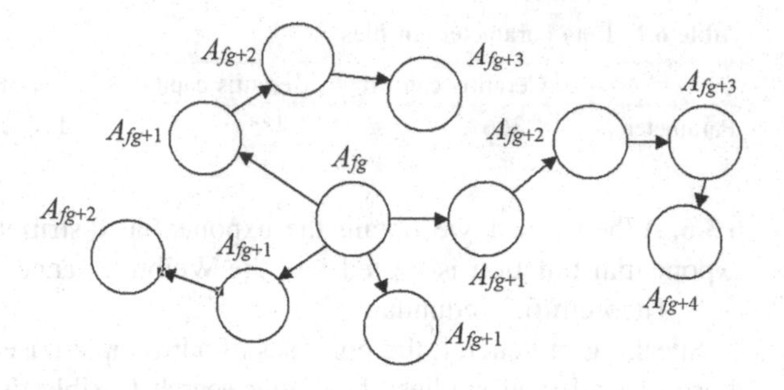

Fig. 6.8 The compound cascade effect centered on A_{fg} and moving toward four directions

6.4.2. The generic sub-state A_{fg} ($g = 1, 2, 3\ldots, n$) degenerates and accelerates the decay of the elements placed all around it. For the sake of simplicity, we subdivide this intricate mechanism into a series of linear waterfall effects which begin with A_{fg} and have different lengths as in Eq. (6.14) (see also Fig. 6.8). In substance, we suppose that the cascade on A_{fg} consists of n degeneration processes that range from one to n steps. The smallest chain contains solely A_{fg}, the largest one includes all the sub-states. Generically speaking, it may be said that A_{fg} causes a two-step chain toward a direction, a three-step chain toward another direction, and so forth. Every part of S decays with time and in principle the previous scheme is valid for every sub-state of A_f.

6.4.3. Now we calculate the hazard rate depending on the *complex cascade hypothesis.*

Theorem 6.4 *Let Eqs.* (6.4) *and* (6.13) *true, the reliability entropy of the functioning state* A_f *which is affected by the complex cascade effect is the following*

$$H_f^+ = -\sum_{k=1}^{n}\left(\frac{|H_{fk}|}{1!} + \frac{|H_{fk}H_{fh}|}{2!} + \frac{|H_{fk}H_{fl}H_{fd}|}{3!} + \cdots + \frac{|H_{fk}H_{fi}H_{fs}H_{fu}\ldots|}{n!}\right). \tag{6.15}$$

Proof As first we calculate the function H_{fk}^+ for the generic sub-state A_{fg} ($g = 1, 2, \ldots n$) which causes n chains of degeneration including the limit chain that contains only g. We use Eq. (6.4) to calculate the entropy of each chain and the sum Eq. (2.15) to get the overall result

$$H_{fk}^+ = -\left(\frac{|H_{fk}|}{1!} + \frac{|H_{fk}H_{fh}|}{2!} + \frac{|H_{fk}H_{fl}H_{fd}|}{3!} + \cdots + \frac{|H_{fk}H_{fi}H_{fs}H_{fu}\ldots|}{n!}\right). \tag{6.16}$$

Equation (6.16) is valid for every part A_{fg} ($g = 1, 2, \ldots n$) thus we calculate the reliability entropy through summation and we prove Eq. (6.15) is true

$$H_f^+ = \sum_{k=1}^{n} H_{fk}^+$$
$$= -\sum_{k=1}^{n}\left(\frac{|H_{fk}|}{1!} + \frac{|H_{fk}H_{fh}|}{2!} + \frac{|H_{fk}H_{fl}H_{fd}|}{3!} + \cdots + \frac{|H_{fk}H_{fi}H_{fs}H_{fu}\ldots|}{n!}\right). \tag{6.17}$$

Theorem 6.5 *Let Eqs.* (4.2) *and* (6.15) *be true, the probability of functioning of S under the compound cascade effect is the exponential-exponential function*

$$P_f = P_f(t) = g\,\exp[-d\,\exp(t)], \quad g, d > 1. \tag{6.18}$$

Proof We make explicit the entropies of Eq. (6.15) using Eq. (4.2)

$$H_f^+(t) = -\sum_{k=1}^{n}\left(\frac{c_k t}{1!} + \frac{c_k c_h t^2}{2!} + \frac{c_k c_l c_d t^3}{3!} + \frac{c_k c_i c_s c_u t^4}{4!} + \cdots\right). \tag{6.19}$$

For the sake of simplicity, we put each hazard rate: c_k, c_h, c_d ... equals to c

$$H_f^+(t) = -n\left(\frac{c\,t}{1!} + \frac{ct^2}{2!} + \frac{c\,t^3}{3!} + \frac{c\,t^4}{4!} + \cdots + \frac{c\,t^n}{n!}\right)$$
$$= -nc\left(\frac{t}{1!} + \frac{t^2}{2!} + \frac{t^3}{3!} + \frac{t^4}{4!} + \cdots + \frac{t^n}{n!}\right). \tag{6.20}$$

It is reasonable to expect that the number of sub-states n is 'very high', hence Eq. (6.20) may be approximated to the exponential series

$$H_f^+(t) = -n\left(\sum_{j=1}^{\infty}\frac{t^j}{j!}\right) = -d\,\exp(t), \quad d > 1. \tag{6.21}$$

By definition, the entropy H_f^+ is the logarithm of P_f

$$H_f^+(t) = \ln(P_f) = -d\,\exp(t).$$

Thus

$$P_f(t) = g\,\exp[-d\,\exp(t)].$$

Equation (6.18) is proved where g is a constant depending on the system.

Lemma 6.2 *From definition* (1.6) *and* (6.18) *we get*

$$\lambda(t) = -\frac{P'(t)}{P(t)} = -\frac{(-dg)\exp[t - d\,\exp(t)]}{g\,\exp[-d\,\exp(t)]} = d\,\exp(t), \quad d > 1. \tag{6.22}$$

The probability of good functioning for a system affected by the complex cascade effect complies with the exponential-exponential law; and the hazard rate is exponential of time.

6.5 Comments on Theorem 6.5 and Lemma 6.2

Living beings have very intricate structures and the complex cascade effect is typical of those systems.

6.5.1. A cell—the brick of any vegetable and animal being—includes *over two hundred thousand chemical components* and this number should give an idea of the intricate and multiple relationships running inside a living being.

Compound cascade degeneration can occur at any level of the hierarchical model and even crosses the structure in the vertical direction. For instance, *genetic diseases* which originate from a change of the genome and affect a variety of cellular functions begin to harm the bottom level in Eq. (2.4) and later injure the upper levels.

$$\begin{aligned} \textit{Human Body} &= \{\text{Alive}\}OR\{\text{Dead}\} \\ &= \{[\text{System}_A]AND[\text{System}_B]AND[\text{System}_C]AND\ldots\}OR\{\ldots\} \\ &= \{[(\text{Organ}_{A1})AND\left(\text{Organ}_{A2}\right)AND\ldots\Big]AND[\ldots]\ldots\}OR\{\ldots\} \\ &= \{[(\text{Tissue}_{A11}AND\ \text{Tissue}_{A12}\ AND\ldots)AND(\text{Tissue}_{A21}\ AND\ldots)\ldots]AND[\ldots]\ldots\}OR\{\ldots\} \\ &= \{[((\text{Cell}_{A111}AND\ \text{Cell}_{A112}AND\ldots)AND(\text{Cell}_{A121}AND\ \text{Cell}_{A122}AND\ldots)\ldots)\ldots]AND[\ldots]\ldots\}OR\{\ldots\} \end{aligned} \tag{2.4}$$

6.5.2. Common diseases also present a set of symptoms, such as fever, tiredness, and pains, which demonstrate the global result of multidirectional actions taking place in the human body, i.e., the flu hits various interdependent organs such as the head, the stomach, the legs and others [12]. A defective function or a disturbance can trigger an intricate cascade of biochemical events.

6.5.3. We pinpoint that tiny degenerative effects can be easily examined. In addition, an organ of the living system adjusts itself when it is injured. It tends towards self-healing and the global system deviates from the definition of *conservative systems* presented in Sect. 3.4. When a self-adjusting process begins, it enfeebles the chained damages forecast by Eq. (6.13). Sometimes one cannot observe the compound cascade effect because of this homeostatic feedback mechanism. Hence we pay attention to the most acute debilitating diseases such as the *multi-organ failure* (MOF), which offers better examples of how a compound cascade effect can occur.

MOF is defined as a clinical syndrome characterized by the development of a progressive physiologic dysfunction in many organs at the upper and lower levels

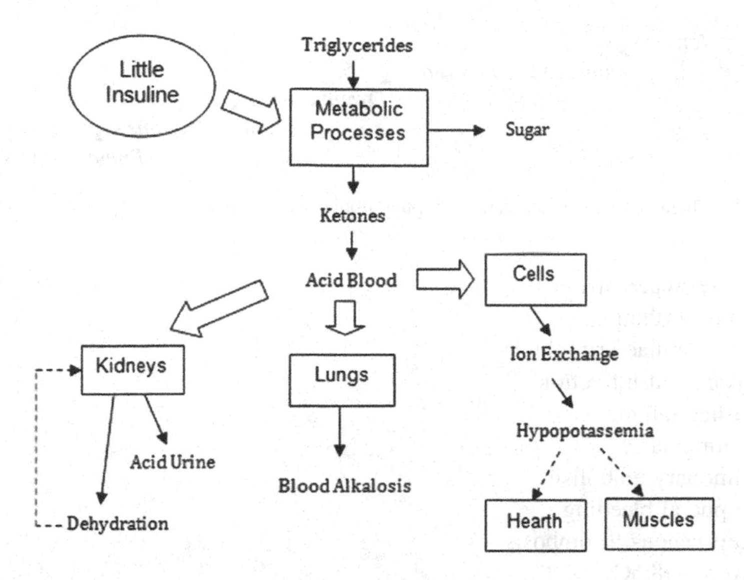

Fig. 6.9 Complex cascade effect of diabetic ketoacidosis (partial) Legenda: → input/output; --▸ damaging action; ⇒ trigger

in Eq. (2.4) which is induced by a variety of acute insults; examples are *ictus, sepsis, Ebola* and *diabetes*.

Example Diabetes is a MOF which usually lasts several years and has the potential to damage various organs over time. When a patient is affected by diabetes, the metabolic processes input triglycerides and output ketones in the blood as a result of the low level of insulin (Fig. 6.9). Blood acidosis interferes with lungs, kidneys and the cell metabolism that carries about insufficient potassium and in turn can damage the hearth and other muscles. The reader can note how the various chains of degeneration include trees and a loop. Various chains reach distant organs such as the muscles of the legs.

Compound damage is evident as a consequence of a *degenerative disease*; examples are *Alzheimer's disease, osteoarthritis, chronic traumatic encephalopathy, macular degeneration, rheumatoid arthritis* and *multiple sclerosis.* These diseases are more frequent in old individuals and this preference is consistent with assumption **(c)** in 6.1.1.

Example Weimar and others [13] make an inquiry concerning the complications following an ictus (or ischemic stroke). They make an itemized report of the following syndromes.

1. Increased intracranial pressure
2. Recurrent cerebral ischemia
3. Intracerebral hemorrhage
4. Epileptic seizure

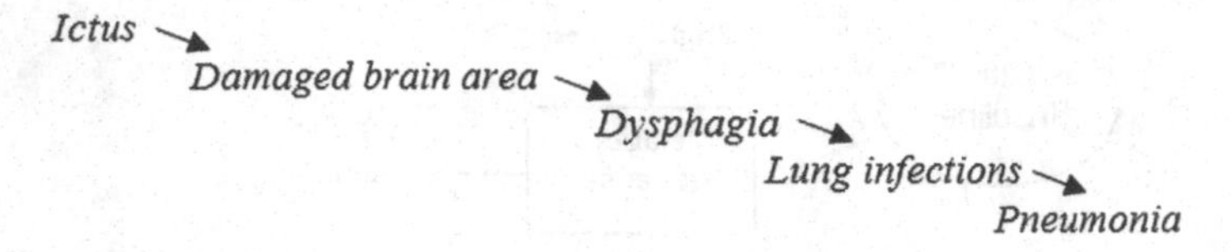

Fig. 6.10 Chained faults which cause the pneumonia after an ictus

5. Severe hypertension
6. Atrial fibrillation
7. Other cardiac arrhythmia
8. Myocardial infarction
9. Cardiac failure
10. Pneumonia
11. Pulmonary embolism
12. Peripheral bleeding
13. Deep venous thrombosis
14. Fever > 38 °C
15. Urinary infection.

These connected diseases affect the brain, the lungs, the heart, the kidneys, the bladder, the circulatory system, and the immune system. Each complication spells out how various linear cascade effects progress toward different directions and involve a small or large number of organs. For instance, the increased intracranial pressure (complication #1) is a local syndrome. It is close to the initial ictus and could be classified as a short damage chain. The cascade effect resulting in pneumonia (complication #10) turns out to be rather complicated (Fig. 6.10). The ischemic stroke damages the area of the brain that controls the throat and this injury results in dysphagia. The patient has arduous swallowing and coughing. He inhales the bacteria flying in the air but cannot eject them. The patient does not expectorate and does not succeed in protecting himself against pulmonary infections, eventually he develops pneumonia.

6.5.4. The British actuary Benjamin Gompertz proposed a probability model for human mortality, based on the assumption that the "*average exhaustion of a man's power to avoid death to be such that at the end of equal infinitely small intervals of time he lost equal portions of his remaining power to oppose destruction which he had at the commencement of these intervals*" [14]. He formalized the statistical distribution which bears his name and published it in the Philosophical Transactions of the Royal Society of London in 1825.

The Gompertz distribution complies with result Eq. (6.18) when $\beta > 1$

$$f(t; \alpha, \beta) = \alpha\beta \exp[-\beta \exp(\alpha t)], \quad t > 0, \alpha > 0, \beta > 1. \tag{6.23}$$

where α and β are the *scale* and *shape* parameters in the order. The mortality rate fits with Eq. (6.22)

$$\lambda(t) = -\alpha\beta \exp(\alpha t), \quad t > 0, \alpha > 0, \beta > 1.$$

6.5.5. Modern writers agree on the following points:

(c1) Complexity is a typical feature of living beings especially the human;
(c2) The Gompertz distribution provides good account of the mortality of people and is normally used in life insurance, demography, epidemiology, actuarial science etc. as a standard mathematical tool [15].

6.5.6. The previous pages develop the following inference

Linearity Assumption + Complex Cascade Effect

$\Downarrow$

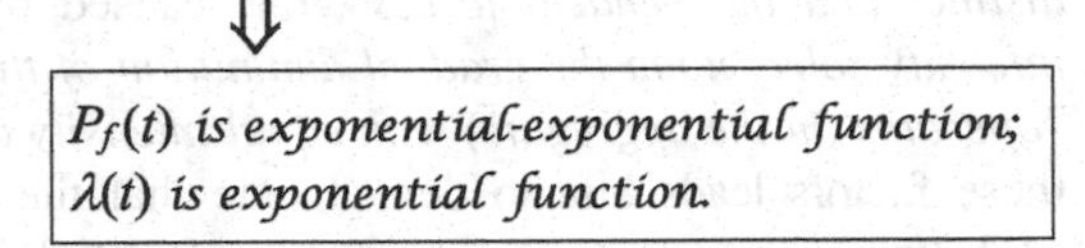

$P_f(t)$ is exponential-exponential function;
$\lambda(t)$ is exponential function.

This logical implication is consistent with the current literature and in addition demonstrates how (c2) depends on (c1). Theorem 6.5 proves that the reliability of S is an exponential-exponential of time since the cascade effect works in an intricate manner.

6.5.7. A lot of empirical investigations validate the exponential-exponential law.

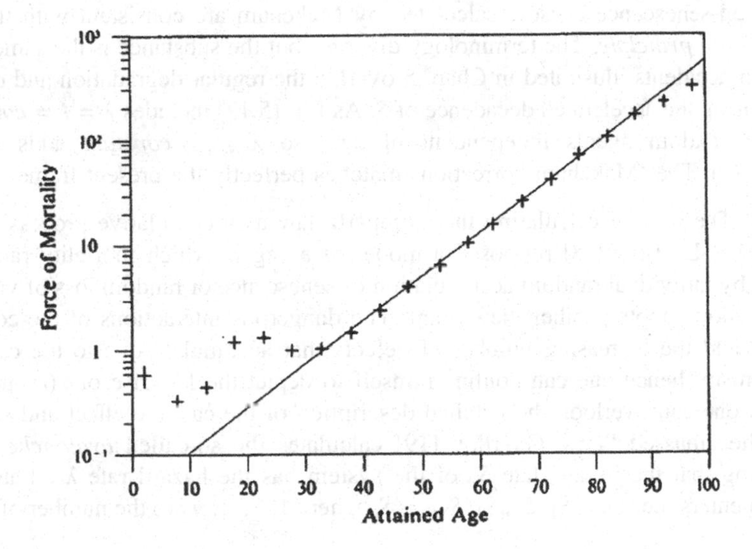

Fig. 6.11 The mortality force of a human cohort by age. Reproduced from [16] Copyright © 1981 by the Society of Actuaries, Schaumburg, Illinois. Reprinted with permission

Example Raw data collected by Wetterstrand [16] shows how the force of mortality is exponential approximately between 30 years of age and 80 years and this is consistent with the compound cascade effect calculated by Eq. 6.22 (Fig. 6.11). The diagram indicates that the mortality rate decelerates in the late lifespan and we shall discuss this phenomenon in Appendix C.

6.5.8. One can define the Gompertz distribution as a log-Weibull distribution from a purely statistical viewpoint. However, Theorems 4.1, 6.3 and 6.5 are not redundant in that their results derive from three distinct sets of hypotheses which describe different mechanisms for the system degeneration. Each hazard rate function has its own precise physical origin.

6.5.9. William Makeham puts forward the distinction between *senescent* and *non-senescent mortality*. He divides the total force of mortality for adults into "*two distinct and independent forces*", one caused by "*diseases depending for their intensity solely upon the gradual diminution of the vital power*" and the other that "*operates (in the aggregate) with equal intensity at all ages*". The superposition of these factors leads him to demonstrate that the addition of the constant ξ—*the Makeham constant*—to the original exponential distribution Eq. (6.23) provides a better fit to the observed age specific mortality rates than to the Gompertz model alone [17]

$$f(t; \alpha, \beta) = \alpha\beta \exp -[\xi t + \beta \exp(\alpha t)], \quad t > 0, \alpha > 0, \beta, \xi > 1.$$

The theoretical mortality rate has this shape

$$\lambda(t) = -[\xi + \alpha\beta \exp(\alpha t)], \quad t > 0, \alpha > 0, \beta, \xi > 1. \tag{6.24}$$

The 'non-senescence causes' calculated by Makeham are consistent with the *superposition principle.* The terminology diverges but the substance is the same. The random accidents illustrated in Chap. 5 overlap the regular degradation and do not stop during the accelerated decadence of S. As Eq. (5.17) includes $\lambda = f =$ *constant* due to random agents independent of age, so $\lambda = \xi =$ *constant* adds up to Eq. (6.24). The 'Makeham correction' matches perfectly the present frame.

6.5.10. The idea of calculating the Gompertz law as a cumulative process is not new. First Le Bras [18] proposed a model of aging in which mortality rates are driven by individual random accumulation of senescence or random loss of vitality.

Gavrilov adopts another view point. The dangerous interactions of the components yield the increasing number of defects that accumulate due to the cascade mechanism; hence one can confine himself to depict the fast rise of $\lambda(t)$. In substance, one can overlook the detailed description of the cascade effect and studies only the amassed flaws. Gavrilov [19] calculates the so-called *avalanche effect* assuming that the initial state S_0 of the system has the hazard rate λ_0. Later the system enters the states S_1, S_2, S_3, S_3, … S_n where 1, 2, … n are the number of flaws which qualify the sequel of states. Gavrilov describes the amass of defects using this arithmetic series

$$\begin{aligned} &\lambda_0; \\ &\lambda_1 = \lambda_0 + \lambda; \\ &\lambda_2 = \lambda_0 + 2\lambda; \\ &\lambda_3 = \lambda_0 + 3\lambda; \\ &\vdots \\ &\lambda_n = \lambda_0 + n\lambda. \end{aligned} \tag{6.25}$$

The multiplicative growth of defects leads to a system of differential equations from which Gavrilov derives the theoretical justification of the Gomperts-Makeham law. In conclusion, Gavrilov disregards the root causes of Eq. (6.25)—that is, the interaction of the parts typical of Eq. (6.23)—and calculates the outcomes of this multiplicative effect, which are the accumulated damages. It is astonishing that both the *avalanche theory* of Gavrilov and the *cascade theory* used in the present book yield identical conclusions.

Further remarks on the Gavrilov theory of aging and longevity will follow in Appendix C.

6.6 Additional Remarks on the Constant and Accelerated Decline of Systems

Let us examine further aspects of the accelerated decline of systems.

6.6.1. All the results obtained in the first part of the book conform to the Gnedenko lesson which proves how chained operations result in the exponential function of the reliability $P_f(t)$. In fact, every system model that has been employed here is a special Markovian system.

6.6.2. The *linearity hypothesis* (4.1) establishes that the rate of entropy change over time is constant

$$\frac{\Delta H_{fg}}{\Delta t} = \frac{[H_{fg}(t) - H_{fg0}]}{(t - t_0)} = \frac{[H_{fg}(P(t)) - H_{fg}(P(t_0))]}{(t - t_0)} = -\, c_g.$$

Time t_0 is the initial instant of the decay process under examination and in principle it can be any. Time t_0 can coincide with whatsoever moment of the system's life span. Even extreme shocks that follow the binomial distribution can occur anytime. The hypotheses of all the theorems and lemmas proved in this part of the book do not depend on the initial value of time but on the changes ΔH_{fg} and Δt.

In summary, this theoretical frame proves that the constant degradation leads to the constant $\lambda(t)$; the linear and the complex cascade effects fix the power and the exponential trends of $\lambda(t)$ in the order. Finally, the reduction of the system structure demonstrates the mortality plateau (Appendix C). In principle, *each hazard rate function does not depend on* t_0 *but can occur in any period of the system lifetime.* This remark is preliminary to the concepts that will be discussed in the next chapter.

6.6.3. There is great concern with respect to the dynamic behaviors of populations that operate in various contexts, including:

- *biological collectives* e.g., insect colonies, bird flocking, bacteria swarming and cells in tissues [20];
- *human groups* e.g., pedestrian and automobile traffic [21];
- *technological elements* e.g., molecular particles, robotic systems and sensor networks [22, 23].

The dynamic behaviors of the crowd emerge from local and simple interactions e.g., attraction and repulsion. Decisions are made independently by each member of the group and can generate, at a macroscopic level, complex behaviors that surprisingly exhibit forms of self-organization. Each member makes decisions through local information transfers between members that are mediated by one or more signaling mechanisms. Researchers adopt a broad assortment of mathematical techniques to calculate global phenomena on the basis of singular behaviors. Perhaps *cellular automata*—implemented with computability techniques—are the most intriguing and innovative models from the present perspective [24].

As regard this book, we recall that until recently, the vast majority of dynamic models of collective behaviors relied on the basic assumption that each individual's neighborhood of interaction was dependent on a metric distance corresponding to the fixed range of its sensory capabilities. More recent evidence [25] shows how interactions amongst the members are mostly metric free, as each member reacts primarily with a limited number of their nearest neighbors, irrespective of the distances between them. This fact has motivated the concept of '*topological interaction*', which is the range measured in units of members, rather than meters. The topological distance d_T is usually the average number of neighbors each member is interacting with. This criterion involves also the present study. We have observed how a degenerated component of S affects the closest component, hence the topological distance is the unit in the linear cascade effect

$$d_T = 1$$

Instead, a single element can interact with $(n - 1)$ elements during the complex cascade effect

$$d_T = (n - 1)$$

6.6.4. It is worth mentioning that one can get the exponential-exponential function following a mathematical approach that differs from Theorems 6.4 and 6.5. Tsitsiashvili almost obtains the accuracy rate of convergence of the minimization scheme to the exponential-exponential distribution [see Appendix B]. Specifically, he adopts the method by Zolotarev [26] who explains how it is convenient to represent some limit distributions by means of maximization or minimization methods and how to estimate the rate of convergence to these limit distributions.

References

1. Yoshida, J., Chino, K., & Wakita, K. (1982). Degradation behavior of AlGaAs double-heterostructure laser diodes aged under pulsed operating conditions. *IEEE Journal of Quantum Electronics, 18*(5), 879–884.
2. Wang, Y., & Bierwagen, G. P. (2009). A new acceleration factor for the testing of corrosion protective coatings: Flow-induced coating degradation. *Journal of Coatings Technology and Research, 6*(4), 429–436.
3. Samuel, M. P. (2014). Differential degradation assessment of helicopter engines operated in marine environment. *Defence Science Journal, 64*(4), 371–377.
4. Kosec, B., Nagode, A., Budak, I., & Karpe, B. (2010). *Sankey diagram samples*. University of Ljubljana Press.
5. Bond, J. Q., Upadhye, A. A., Olcay, H., Tompsett, G. A., Jae, J., Xing, R., et al. (2014). Production of renewable jet fuel range alkanes and commodity chemicals from integrated catalytic processing of biomass. *Energy and Environment Science, 7,* 1500–1523.
6. Barlow, R. E., & Proschan, F. (1975). *Statistical theory of reliability and life testing*. NY: Holt, Rinehart and Winston.
7. Rigdon, S. E., Basu, & A. P. (2000). *Statistical methods for the reliability of repairable systems*. Wiley.
8. Sheikh, A. K., Younas, M., & Al-Anazi, D. M. (2002). Weibull analysis of time between failures of pumps used in an oil refinery. *Proceedings of 6th Saudi Engineering Conference, 4*, 475–491.
9. Janse, C., Slob, W., Popelier, C. M., & Vogelaar, J. W. (1988). Survival characteristics of the mollusc Lymnaea stagnalis under constant culture conditions: Effects of aging and disease. *Mechanisms of Ageing and Development, 42*, 263–174.
10. Hirsch, A. G., Williams, R. J., & Mehl, P. (1994). Kinetics of medfly mortality. *Experimental Gerontology, 29,* 197–204.
11. Duyck, P. F., David, P., & Quilici, S. (2007). Can more K-selected species be better invaders? A case study of fruit flies in La Réunion. *Diversity and Distributions, 13,* 535–543.
12. Deyo, R. A. (2002). Cascade effects of medical technology. *Annual Review Public Health, 23,* 23–44.
13. Weimar, C., Roth, M. P., Zillessen, G., Glahn, J., Wimmer, M. L. J., Busse, O., et al. (2002). Complications following acute ischemic stroke. *European Neurology, 48,* 133–140.
14. Johnson N. L., Kotz, S., & Balakrishnan, N. (1995). *Continuous univariate distributions, 2*. Wiley.
15. Olshansky, S. J., & Carnes, B. A. (1997). Ever since Gompertz. *Demography, 34*(1), 1–15.
16. Wetterstrand, W. H. (1981). Parametric models for life insurance mortality data: Gompertz's law over time. *Transactions of the Society of Actuaries, 33*, 159–175.
17. Bongaarts, J. (2009). Trends in senescent life expectancy. *Population Studies, 63*(3), 203–213.
18. Le Bras, H. (1976). Lois de mortalité et age limite. *Population, 33*(3), 655–691.

19. Gavrilov, L. A., & Gavrilova, N. S. (1991). *The biology of life span: A quantitative approach.* Harwood Academic Publisher.
20. Couzin, I. D., Krause, J., James, R., Ruxton, G. D., & Franks, N. R. (2002). Collective memory and spatial sorting in animal groups. *J. of Theoretical Biology, 218*(1), 1–11.
21. Nagel, K. (1996). Particle hopping models and traffic flow theory. *Physical review E, 53,* 4655–4672.
22. Leonard, N. E., Paley, D. A., Lekien, F., Sepulchre, R., Fratantoni, D. M., & Davis, R. E. (2007). Collective motion, sensor networks, and ocean sampling. *Proceedings of the IEEE, 95*, 48–74.
23. Blanchet, A., & Degond, P. (2016). Topological interactions in a Boltzmann-Type framework. *Journal of Statistical Physics, 163,* 41–60.
24. Chopard, B., & Droz, M. (2005). *Cellular automata modeling of physical systems.* Cambridge University Press.
25. Ballerini, M., Cabibbo, N., Candelier, R., Cavagna, A., Cisbani, E., & Giardina, I. (2008). Interaction ruling animal collective behaviour depends on topological rather than metric distance: Evidence from a field study. *Proceedings of the National Academy of Sciences USA, 105*, 1232–1237.
26. Zolotarev, V. M. (1997). *Modern theory of summation of random variables.* VSP International Science Publishers.

Chapter 7
When the Premises Are Untrue

A principle-based theory provides the results through deductions of this kind

$$\mathcal{A} \Rightarrow \mathcal{B} \tag{7.1}$$

In the previous pages we have reasoned under the assumption that the premises were true, but this is not enough. One cannot accept the "one size fits all" approach. The careful search for the hazard rate functions should also examine the situations when the premises are false and should infer consistent outcomes from these hypotheses. The inverse of (7.1) yields

$$\text{Not-}\mathcal{A} \Rightarrow \text{Not-}\mathcal{B} \tag{7.2}$$

Classic logic teaches us that if the premise is false, then the conclusion is also false. We mean to examine various critical cases where the calculations of the previous pages are not applicable. The main premises $\mathcal{A}$ of the present frame include the entropy function, the *linearity assumption*, the cascade effect and other hypotheses. We are going to analyze the negation of a single hypothesis or even the contemporary negation of two or more hypotheses.

7.1 The Entropy Does not Exist

The Boltzmann-like entropy is proved under the axiom (**3**) in Sect. 2.3.3. This basic constraint holds that the I/R property of the macro-state A_i depends on the I/R property of the micro-states. Instead various works show how the reliability of the overall system may not rely on the quality of the parts as indicated in the historical paper of Moore and Shannon [1]. Ascher claims that sometimes there is little connection between the properties of the component hazard rates and the properties of the overall process [2]. In the cited cases, axiom (**3**) is false since the I/R property

P. Rocchi, *Reliability is a New Science*, DOI 10.1007/978-3-319-57472-1_7

of S should depend on the I/R of its parts. Theorem 2.1 is false and one cannot calculate the entropy H_r.

Example Elerath and Pecht [3] claim that the reliability of an array of disks is not related to the reliability of a single hard disk drive. That is, the times between successive system failures can become increasingly longer even though each component hazard rate is increasing. One can apply the present theory to the single hard disk drive but not to the overall array of disks since the entropy function H_r is false.

7.2 The Linearity Assumption Is False

The *principle of progressive deterioration* is universal but the manner in which it is applied can vary. Specifically, the most common exceptions regard the *linearity assumption*. In Sect. 4.2.3 we showed how the degradation function $D_g(t)$ can follows different trends and symmetrically even the function $H_{fg}(t)$ can be any.

Let us study the components whose entropy does not diminish with time.

7.2.1 The parts of a very robust system keep their qualities for a long while. One can reasonably assume that the effectiveness is constant

$$H_{fg}(t) = H_{fg} = -q, \quad g = 1, 2, \ldots, n, \quad q > 0. \tag{7.3}$$

The reliability entropy H_f of S is the summation of partial entropies and we get the probability of good functioning

$$H_f(t) = \sum_{g=1}^{n} H_{fg} = -\sum_{g=1}^{n} q = -nq,$$
$$\log P_f(t) = -nq,$$
$$P_f(t) = \exp(-nq).$$

Finally

$$\lambda(t) = -\frac{P'(t)}{P(t)} = -\frac{0}{\exp(nq)} = 0, \quad q > 0, g = 1, \ldots, n. \tag{7.4}$$

In practice, $\lambda(t)$ tends to zero when (7.3) is true such as in the next example regarding living beings affected by the so-called *negligible senescence*.

Example Biologists have found that some living beings do not experience a measurable functional decline with age. For example, some organisms of this group have no post-mitotic cells; or they reduce the effect of damaging free radicals by cell division and dilution, or they even have apparently limitless telomere

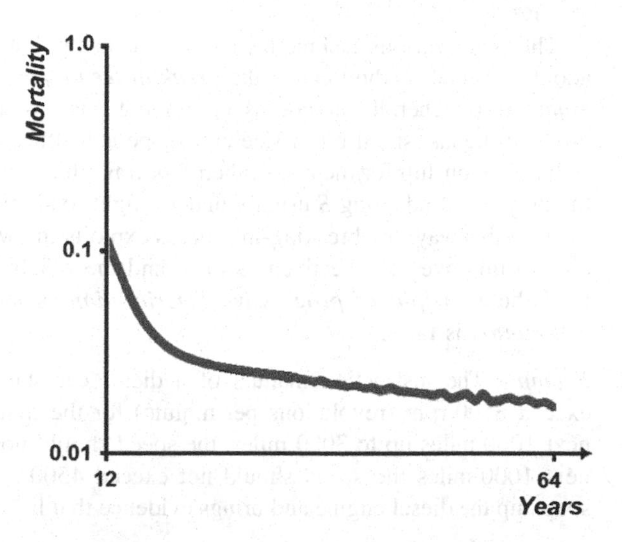

Fig. 7.1 Mortality of Gopherus agassizii (desert tortoise). Modified from [7] by permission from Macmillan Publishers Ltd

regenerative capacity. Negligibly senescent organisms exhibit good performances even if old and the death rates do not increase with age in a significant manner [4–6]. Jones [7] have found animal and plant species whose mortality gets close to zero in accordance with (7.4) (Fig. 7.1).

Vaupel and others [8] argue that negligible senescence may not be as theoretically implausible as some had supposed, and that it might not even be so rare [9].

7.2.2 There are components whose capability of working improve with time, the reliability entropy climbs up and does not diminish as stated in the *linearity assumption* (4.2). For the sake of simplicity, let H_{fg} augments as a square of time and this rule is the same for each sub-state of A_f

$$H_{fg}(t) = qt^2, \quad q > 0 \quad g = 1, 2, \ldots, n. \tag{7.5}$$

We obtain the entropy H_f and in turn the probability of good functioning of S

$$H_f(t) = \sum_{g=1}^{n} H_{fg} = \sum_{g=1}^{n} qt^2 = nq \cdot t^2,$$
$$P_f(t) = \exp(nqt^2).$$

The hazard function diminishes with time

$$\lambda(t) = -\frac{P'(t)}{P(t)} = -\frac{nqt\,\exp(nqt^2)}{[\exp(nqt^2)]} = -nqt + r \quad q, r > 0 \tag{7.6}$$

where r is the value of $\lambda(t)$ for $t = 0$.

There are various techniques and mechanisms that apply (7.6). Manufacturers adopt a special method called the *break-in technique*, also known as *run-in technique*. It is generally a process of moving parts wearing against each other to produce the last small bit of size and shape adjustment that will settle them into a stable relationship for the rest of their working life. This training procedure pursues the purpose of adapting S usually under a light load, but sometimes under a heavy load. In this way, the breaking-in process exploits the wear effect in a manner to be able to improve the effectiveness of S, and the reliability entropy increases. Note how the *principle of progressive deterioration* is still true while the *linearity assumption* is false.

Example The instruction manual of a diesel car states that the speed must not exceed 3500 rpm (revolutions per minute) for the first 1000–2000 miles. For the next 1000 miles up to 3000 miles, the speed should not exceed 4000 rpm. For the next 1000 miles the speed should not exceed 4500 rpm. The break-in technique soups up the diesel engine and brings evidence that linear decline is false (Fig. 7.2).

Infant mortality refers to deaths of children. Many factors contribute to infant mortality: medical causes, the environment, the socio-economic situation, wars, and the cultural level. Joined agents result in a high mortality rate during the first years, which progressively diminishes with time. Several works have analyzed the mortality rate during infantry [10].

Example We confine ourselves to mentioning the survey conducted in Italy from 1999 to 2001 [11] (Fig. 7.3).

An animal grows from a one-cell system to an adult being that has well developed organs and behavior. The body's development, which yields the complete maturity of the individual, turns out to be a powerful phenomenon as it undoes the decline involved by the *principle of progressive deterioration.* A body that is growing is capable of inverting the influence of that principle on mortality.

Example The dimension of the seal body (male) rises at a high rate and the mortality decreases during the initial lifetime. The growth rate lessens after the fourth year of life and the mortality rate climbs up during this period. Clinton and Le Boeuf find out that wo opposite trends occur in the interval 1–4 years of age and in the interval 4–14 [12] (Fig. 7.4).

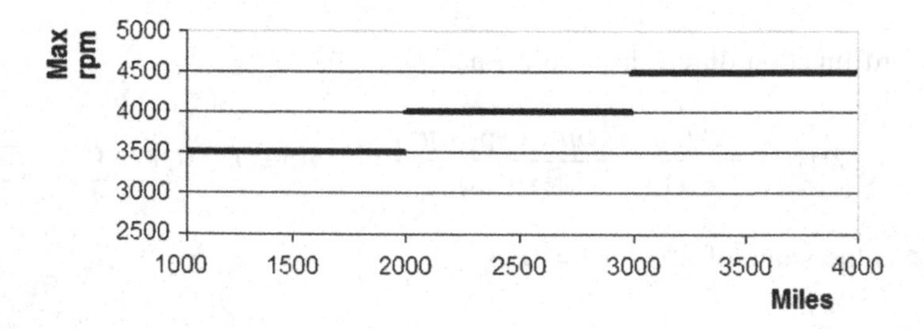

Fig. 7.2 Constant enhancement of diesel engine rotation speed with break-in

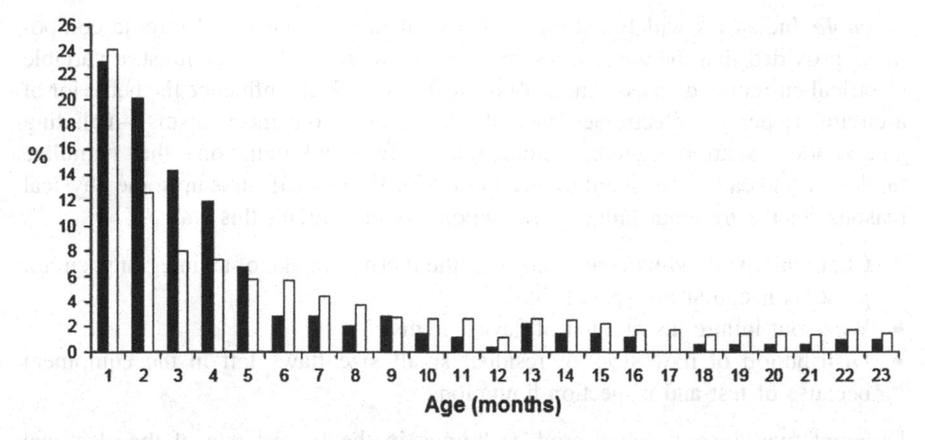

Fig. 7.3 Descending percentage of deaths due to sudden infant death syndromes (*black bars*) and to other causes (*white bars*). Reproduced from [11] with permission from Istituto Superiore di Sanità, Roma

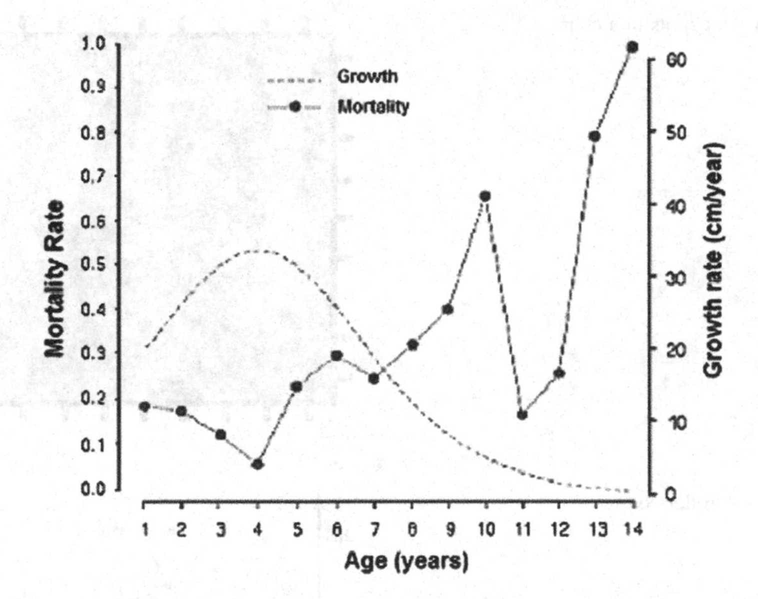

Fig. 7.4 The male growth curve versus the age-specific male mortality curve. Modified from [12] with permission from Wiley

7.2.3 Let us examine when the *hypothesis of linearity* is true but *the sub-states of S do not have a common behavior*, in the sense that the side effects superpose with different weights at different times and the reliability entropy H_{fg} follows an irregular trend. The curve $\lambda(t)$ presents peaks and humps because of asymmetric overlapping factors.

Example Industries widely accept the constant hazard rate for electronic components provided that the constant current crosses them [13]. By contrast, a variable electrical current and an assortment of intrinsic side effects influence the behavior of a circuit. Experts in electronics make the list of wear-out mechanisms—including gate oxide, electromigration, channel hot carriers and radiation—that cumulate randomly and cause the circuit to break [14, 15]. Wong [16] sums up some physical reasons for the irregular failures of components in a circuit this way:

- Changing overloading conditions. E.g. the thermal image of an integrated circuit presents irregular hot spots (Fig. 7.5);
- Wear-out failure distribution of flawed items;
- Distribution of flaw sizes or residual small size flaws left in the equipment because of test and inspection limitations.

Different overlapping agents produce humps in the hazard rate of the electrical circuit (Fig. 7.6) which some writer defines as a *roller-coaster* curve (also Fig. 1.3).

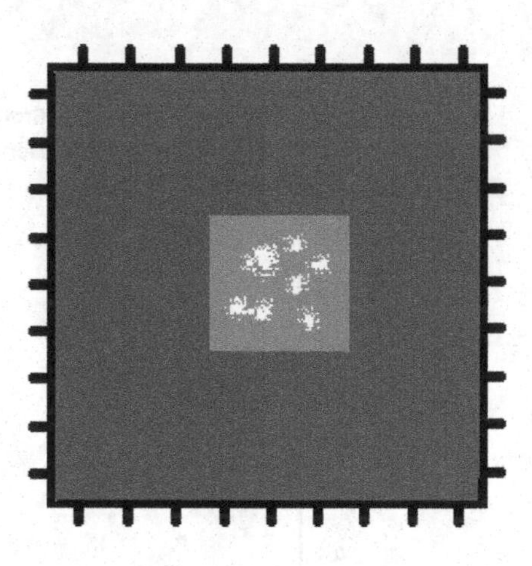

Fig. 7.5 Hot spots in a chip

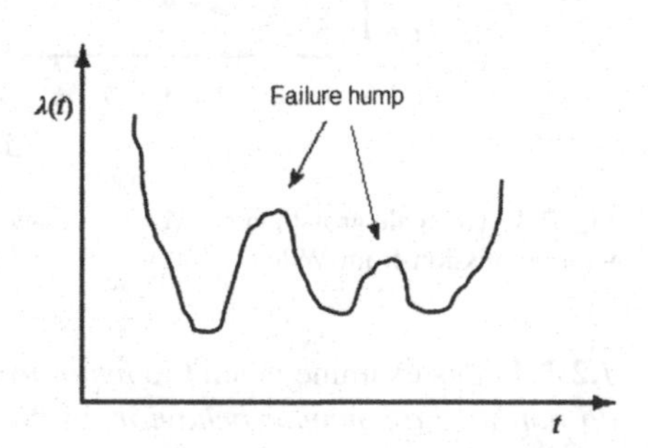

Fig. 7.6 A roller-coaster curve

7.2.4 Scholars assert that aging is an essential and innate part of the biology of living beings. The *theories of programmed aging* state that aging is predetermined in a way by the intervention of biological mechanisms which are connected to the endocrine system, the immune system and the genes in a way. When a programmed mechanism of aging overwhelms the decline processes, the *linearity assumption* is false and as a consequence the organism does not comply with the Gompertz law.

Example When Pacific salmons have lived in the ocean for 2 or 3 years, they make an upstream journey to find a place suitable for spawning. After spawning, the adrenal gland releases massive amounts of corticosteroids that in a few days cause the salmon to die. In a rather short period of time, Pacific salmons undergo a rigid sequence of events consisting of upstream migration, spawning and death [17]. The mortality rate of Pacific salmon does not fit with the results obtained in the previous chapters.

7.3 The Cascade Effect Is False

The cascade effect occurs when a degenerated part spoils a neighboring part and this in turn affects another part and so on. This hypothesis may be false for several reasons.

7.3.1 Sometimes the inner structure of the system is incompatible with the waterfall mechanism.

Example Ceramic items are increasingly being used in cutting edge technologies even if they present some shortcomings. The poor reliability in strength, or the rather large variability in the strength property of ceramics, is largely due to the variability in the distribution of crack sizes, shape and orientation with respect to the tensile loading axis. The rigidity and the frailty of ceramic material are incompatible with the cascade effect. Experimental evidence proves this conclusion is true [18].

7.3.2 Experts on *Weibull analysis* recognize that stable decaying processes during ageing give straight lines of failure data on logarithm plots, while a random mixture of failure modes produce visible cusps, corners and knees. A small deviation indicates different types of failures, different operating environments, poorly manufactured lots, or parts made by different manufacturers, etc. [19, 20]. The irregular mix of failure factors is inconsistent with the linear cascade effect that Lemma 6.1 assumes as true.

7.3.3 The structure of some systems is modified during advanced aging; thus the cascade effect cannot occur or it may even happen that the modified structure masks the cascade effect.

Example Greenwood and Irwin [21] suggest: "The possibility that with advancing age the rate of mortality asymptotes to a finite value". This scientific fact is known now as the *late-life mortality deceleration.* As documented in the book "*Supercentenarians*" [22] human death rates appear to reach a plateau at about 80 years. This phenomenon falsifies the hypothesis of the compound cascade effect and several theories have been put forward to clarify it. Appendix C discusses this argument.

7.3.4 Hypothesis (6.5) entails that the chain has a certain length. If the system includes a few elements, then the cascade effect either is not apparent or does not occur. For example the structure of electrical transformers includes two chains (Fig. 4.3): the horizontal chain has three blocks; the vertical has only two blocks. It is evident how the waterfall effect has little room and the experimental results, which show constant failure rate, fit with the present theoretical frame.

7.3.5 The *cascading failure* is an event in which a failure of one part of an interconnected system leads to a failure in related areas of the system. The initial fault propagates itself to the point that the overall system crashes. This is a common threat in engineering [23], networking [24], industry [25], ecology [26], biology and even in global politics [27]. Cascading failure is consistent with the concept of *cascade effect* in point of logic but an essential physical property divides the former from the latter. Hypothesis (6.1) implies that the *cascade effect* evolves over a certain interval of time. The present frame calculates the continuous influence that every wasted part of S has on the components nearby, whereas the *cascading failure* is usually a rather rapid event. Experts look into dramatic situations that evolve over a short period of time. The reliability entropy of the cascade effect H_f^* is false and in consequence the Weibull law or the Gompertz law turn out to be rather nonsensical.

Example A DC-10 aircraft cashed over Paris, France, in 1974, killing everyone on board. A later investigation into the cause of the crash revealed that a cargo bay door had not been fastened properly. The man most directly responsible for this reputedly could not read English and therefore was not able to realize the instructions for how to properly fasten the door. As the aircraft climbed to about four thousand meters, internal pressure caused the door to give way, and the explosive decompression around the door as it blew off damaged hydraulic controls in the area, which caused the pilots to eventually lose complete control of the aircraft. The overall cascading failure lasted a few minutes.

7.4 Binomial Assumption Is False

We have supposed that random accidents follow the binomial distribution whose parameter is a positive constant.

7.4.1 This hypothesis is false when the environment of S follows a seasonal rhythm. The events that damage the system do not occur randomly but exhibit a precise behavior.

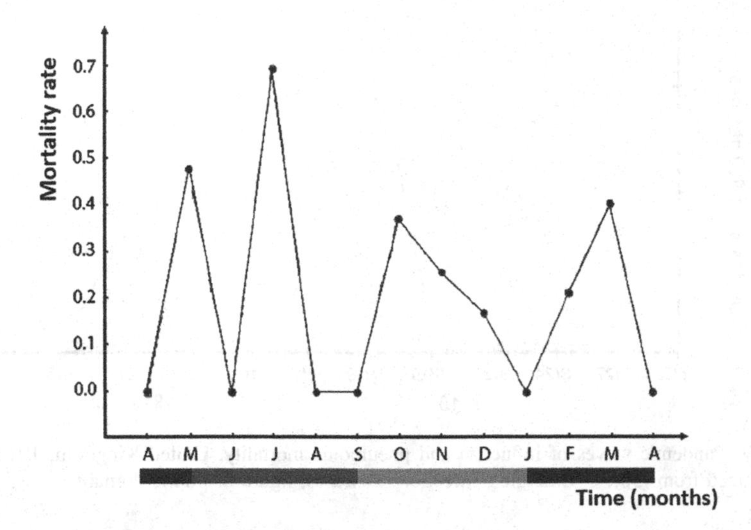

Fig. 7.7 Mortality rate of Zostera marina. Modified from [28] with permission from Revista de Biología Tropical/Journal of Tropical Biology and Conservation, Universidad de Costa Rica

Example The weather influences the development of vegetables and in turn the mortality rate. As an example, take the grasses which grow and die according to a seasonal pattern, such as eelgrasses, whose lifecycle is determined by three seasons: May–August, September–December and January–April [28] (Fig. 7.7. Canfield [29] finds evidence of increased mortality with increasing age in some perennial plants. He shows, however, that mortality varies irregularly depending on the external environmental conditions that may be more or less favorable for survival (Fig. 7.8).

7.4.2 It may happen that a population of animals is attacked by external entities that are able to decimate or even destroy the population. An animal species can suffer from hunger, thirst, or attacks of predators. Humans can become sick and die from

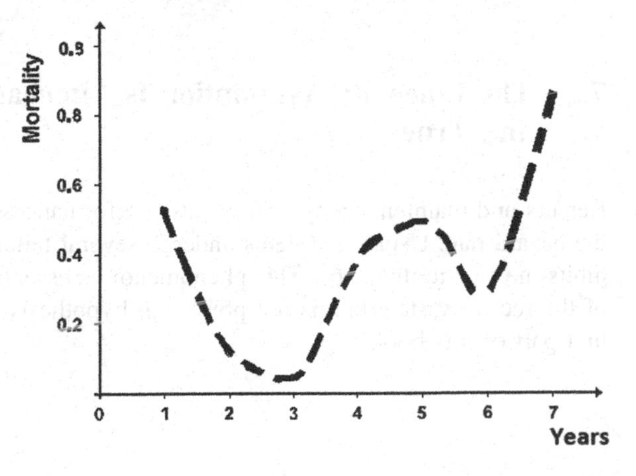

Fig. 7.8 Oscillating trend of mortality for a perennial plant

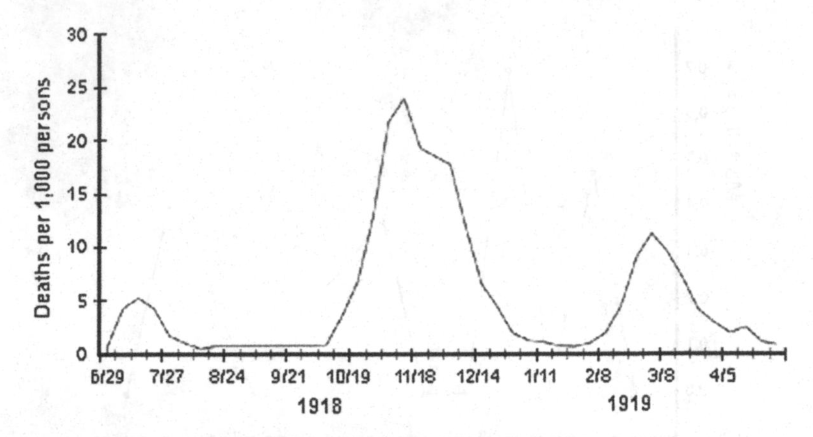

Fig. 7.9 Pandemic waves of influenza and pneumonia mortality, United Kingdom, 1918–1919. Reproduced from [30], © Emerging Infectious Diseases, figure of public domain

massive infections, they can be eliminated by global wars, fierce invaders, or catastrophic environmental events. The *binomial assumption* and the *linearity assumption* are untrue in these situations, and also the cascade effects cannot be attained. Catastrophic events cause the mortality function of the involved populations to deviate from the normal trends.

Examples The *First World War* was a global war centered in Europe that began on 28 July 1914 and lasted until 11 November 1918. Over 9 million combatants and 7 million civilians died as a result of the war including the victims of a number of genocides.

Between January 1918 and December 1920 the so-called *Spanish flu* peaked and infected 500 million people across the world, including remote Pacific islands and the Arctic, and resulted in the deaths of 50–100 million people or about the three to five percent of the world's population. Figure 7.9 offers a visual illustration of the epidemic mortality registered in UK.

7.5 The Linearity Assumption Is Alternatively False and True

Repairs and maintenance typically cause performances to improve and push down the hazard rate. Usually, systems undergo several failure-repair cycles and $\lambda(t)$ exhibits the saw-tooth shape. This phenomenon is essentially due to the intervention of the recovery state that is not present in hypothesis (2.7) which reigns over the first part of this book

$$S = A_f. \tag{2.7}$$

The properties of the recovery state will be amply discussed in the second part of the book.

7.6 Intent on Completeness

The present chapter offers for consideration a dozen cases underpinned by the implication

$$\text{Not-}\mathcal{A} \Rightarrow \text{Not-}\mathcal{B} \tag{7.2}$$

Experimental evidence shows how the hazard rate does not follow the three basic trends—constant, power and exponential—because the assumptions are false. A broad assortment of overlapping failure factors—either continuous or random or both—generate the irregular shapes of $\lambda(t)$. The symbol Not-$\mathcal{A}$ summarizes a variety of agents that frequently are out of control and cannot be theorized.

The detailed discussion of Not-$\mathcal{A} \Rightarrow$ Not-$\mathcal{B}$ aside the regular cases covered by $\mathcal{A} \Rightarrow \mathcal{B}$ show how this book makes a serious effort; it attempts to offer the comprehensive view of the hazard rate function in a variety of situations and environments.

References

1. Moore, F. C., & Shannon, C. E. (1956). Reliable circuits using less reliable relays. *Journal of the Franklin Institute, 262,* part I: 191–208; part II: 281–297.
2. Ascher, H. (1983). Statistical methods in reliability: Discussion. *Technometrics, 25*(4), 320–326.
3. Elerath, J. G., & Pecht, M. (2007). Enhanced reliability modeling of RAID storage systems. In *Proceedings of the 37th IEEE/IFIP International Conference on Dependable Systems and Networks* (pp. 175–184).
4. Rando, T. A. (2006). Stem cells, ageing and the quest for immortality. *Nature, 441,* 1080–1086.
5. Finch, C. E. (1990). *Longevity, senescence and the genome*. University of Chicago Press.
6. Haranghy, L., & Baladázs, A. (1980). Regeneration and rejuvenation of invertebrates. In N. W. Shock (Ed.), *Perspectives in experimental gerontology* (pp. 224–233). Arno Press.
7. Jones, O. R., Scheuerlein, A., Salguero Gomez, R., Camarda, C. G., Schaible, R., Casper, B. B., et al. (2014). Diversity of ageing across the tree of life. *Nature, 505*(7482), 169–173.
8. Vaupel, J. W., Baudisch, A., Dölling, M., Roach, D. A., & Gampe, J. (2004). The case for negative senescence. *Theoretical Population Biology, 65*(4), 339–351.
9. Guerin, J. C. (2004). Emerging area of aging research: Long-lived animals with "negligible senescence". *Annals of the New York Academy of Sciences, 1019,* 518–520.

10. MacDorman, M. F., Mathews, T. J., Mohangoo, A. D., & Zeitlin, J. (2014). International comparisons of infant mortality and related factors: United States and Europe, 2010. *National Vital Statistics Reports, 63*(5), 1–7.
11. Istituto Superiore di Sanità. (2005). Mortalità nei primi due anni di vita in Italia: Sudden Infant Death Syndrome (SIDS) e altre morti inattese - *Rapporti ISTISAN*, 05/2, Ist. Superiore di Sanità Editore.
12. Clinton, W. L., & Le Boeuf, B. J. (1993). Sexual selection's effects on male life history and the pattern of male mortality. *Ecology, 74,* 1884–1892.
13. *Military Handbook, reliability prediction of electronic equipment* (1995), MIL-HDBK-217, Revision F, Notice 2.
14. Nash, F. R. (1993). *Estimating device reliability: Assessment of credibility*. Kluwer Academic Publishers.
15. Evans, J. W., & Evans, J. Y. (2001). Concepts in reliability for design. In J. W. Evans, J. Y. Evans (Eds.), *Product integrity and reliability in design* (pp. 53–88). Springer.
16. Wong, K. L. (1991). The physical basis for the roller-coaster hazard rate curve for electronics. *Quality and Reliability Engineering International, 7*(6), 489–495.
17. McBride, J. R., Fagerlund, U. H. M., Dyel, H. M., & Bagshaw, J. (1986). Changes in structure of tissues and in plasma cortisol during the spawning migration of pink salmon, Oncorhynchus gorbucha (Walbaum). *Journal of Fish Biology, 29*(2), 153–166.
18. Basu, B., Tiwari, D., Kundu, D., & Prasad, R. (2009). Is Weibull distribution the most appropriate statistical strength distribution for brittle materials? *Ceramics International, 35*(1), 237–246.
19. Dodson, B. (2006). *The Weibull analysis handbook*. ASQ Quality Press.
20. Bloch, H. P., & Geitner, F. K. (2012). *Machinery failure analysis and troubleshooting* (4th ed.). Elsevier.
21. Greenwood, M., & Irwin, J. O. (1939). The biostatistics of senility. *Human Biology, 11,* 1–23.
22. Maier, H., Gampe, J., Jeune, B., Vaupe, J. W., & Robine, J. M. (Eds.). (2010). *Supercentenarians.* Springer.
23. Makhankova, A., Barabashc, V., Berkhova, N., Divavina, V., Giniatullina, R., Grigorieva, S., et al. (2001). Investigation of cascade effect failure for tungsten armour. *Fusion Engineering and Design, 56–57,* 337–342.
24. Dorogovtesev, S. N., & Mendes, J. F. F. (2003). *Evolution of networks*. Oxford University Press.
25. Genserik, R. (2009). Man-made domino effect disasters in the chemical industry: The need for integrating safety and security in chemical clusters. *Disaster Advances, 2*(2), 3–5.
26. Smith, K., & Petley, D. (2009). *Environmental hazards. Assessing risk and reducing disaster*. Routledge.
27. Helbing, D. (2013). Globally networked risks and how to respond. *Nature, 497,* 51–59.
28. Uzeta, O. F., Arellano, E. S., & Heras, H. E. (2008). Mortality rate estimation for eelgrass *Zostera marina* (Potamogetonaceae) using projections from Leslie matrices. *Revista de Biologia Tropical, 56*(3), 1015–1022.
29. Canfield, R. H. (1957). Reproduction and life span of some perennial grasses of southern Arizona. *Journal of Range Management, 10*(5), 199–203.
30. Taubenberger, J. K., & Morens, D. M. (2006). 1918 influenza: The mother of all pandemics. *Emerging Infectious Diseases*, *12*(1), 15–22. doi:10.3201/eid1201.050979

Chapter 8
Ideal Hazard Rate

The first chapter of this book recalled for the reader that Gnedenko, Bélyaev and Solovyev [1] approved the bathtub curve after generic considerations. Now we can go back to this endorsement which constitutes a weak point of Gbedenko's seminal work.

8.1 Genericness

8.1.1 Most educators follow the common behaviors of researchers and teach a lesson about the bathtub curve which could be summed up this way:

$$\begin{array}{l}\textit{Most products go through three distinct phases from product}\\ \textit{inception to wear out. In infancy components fail due to defects and}\\ \textit{poor design that cause an item to be legitimately bad. If a component}\\ \textit{does not fail within its infancy, it will generally tend to remain}\\ \textit{trouble -- free over the maturity lifetime. During ageing components}\\ \textit{begin to wear out, and failures start to increase.}\end{array} \tag{8.1}$$

However empirical evidence discussed in Sects. 7.1–7.6 and several scholars—mentioned in Sect. 1.4—show how this verbal description does not match with real data. It may be taken as an intuitive and unproved scheme.

8.1.2 Some are inclined to formalize (8.1) using, for instance, the Weibull distribution with the following constraints:

P. Rocchi, *Reliability is a New Science*, DOI 10.1007/978-3-319-57472-1_8

$$\begin{aligned} &\bullet\ 0<\beta<1 && \text{during infancy;} \\ &\bullet\ \beta=1 && \text{during maturity;} \\ &\bullet\ \beta>1 && \text{during senescence.} \end{aligned} \tag{8.2}$$

Others define the bathtub curve using an analytical function such as Nash [2] who obtains this equation where the parameters are: a, $c > 0$; $b < 1$; $d > 1$ and $t \geq 0$

$$\lambda(t) = ab[at]^{(b-1)} + cd[ct]^{(d-1)}. \tag{8.3}$$

Tuning up the parameters one can obtain the variants of the tripartite curve, but the ways to approximate the statistical distributions are non-trivial.

Concluding, there is no exact support to demonstrate the bathtub curve is true or should be true. The attempts to fix the U-shaped curve using the *inductive* logic have not had full success so far. All that remains is to follow the *deductive* approach and establish $\lambda(t)$ of the whole lifetime on the basis of theorems and lemmas, more precisely, we suggest adopting the following method.

8.2 Deductive Logic and the Tripartite Curve

Section 1.1 introduces the concept of *ideal model* $\mathcal{X}$ which has the purpose of illustrating the visionary or optimal state of something;

8.2.1 Section 1.1 adds up three characteristics of $\mathcal{X}$:

(a) it has to be calculated through deductive logic;
(b) it is usually simple and
(c) precise.

Let us examine the outcomes obtained in the previous chapters respect to the properties listed above:

(b) Lemmas 4.1, 5.1, 6.1 and 6.2 depict three rather regular situations.
(c) Lemmas 4.1, 5.1, 6.1 and 6.2 derive the following equations from rigorous hypotheses:

$$\begin{aligned} &\bullet\ \lambda(t) = \text{const}, \\ &\bullet\ \lambda(t) = t^n, \\ &\bullet\ \lambda(t) = \exp(t). \end{aligned} \tag{8.4}$$

(a) We employ the deductive method and specify the following assumptions:

(a.01) The cascade effect occurs provided that the system components have a certain level of deterioration (assumption (b) in Sect. 6.1) and this hypothesis is appropriate with ageing. Hence it is reasonable to infer that the Weibull and Gompertz laws are appropriate for treating ageing.

(a.02) Section 5.5.2 pinpoints that systematic and random agents are the dominant causes of failure during the entire lifespan, but the cascade effects—typical of the period a.01—mask the actions of CFs and RFs. The cascade effects do not interfere during the system maturity since the components of *S* do not suffer significant deterioration, and one can deduce that the constant hazard rate takes place in the system maturity.

(a.03) Generally, manufacturers' burn-in and infant mortality causes a decreasing $\lambda(t)$ distribution in the first part of the system lifetime.

8.2.2 In summary, one can reasonably assign:

- Power or exponential $\lambda(t)$ to ageing *on the basis of Lemmas* 6.1 or 6.2 *and assumption a*.01;
- Constant $\lambda(t)$ to maturity *on the basis of Lemmas* 4.1, 5.1 *and assumption a*.02;
- Decreasing $\lambda(t)$ to infancy *on the basis of empirical criterion assumption a*.03.

(8.5)

By joining functions (8.5) one can draw the ideal hazard rate which covers the whole system life:

The U-shaped curve is the ideal model of a system's hazard rate. (8.6)

The tripartite curve presents a simplified image of the hazard rate and can be approved as *the finest description of the functioning system's behavior* (Fig. 8.1).

8.3 Comments on the Bathtub Curve

The reader can note how on the one hand descriptions (8.1), (8.2) and (8.3) and on the other hand ideal model (8.5) have far differing backgrounds and contents.

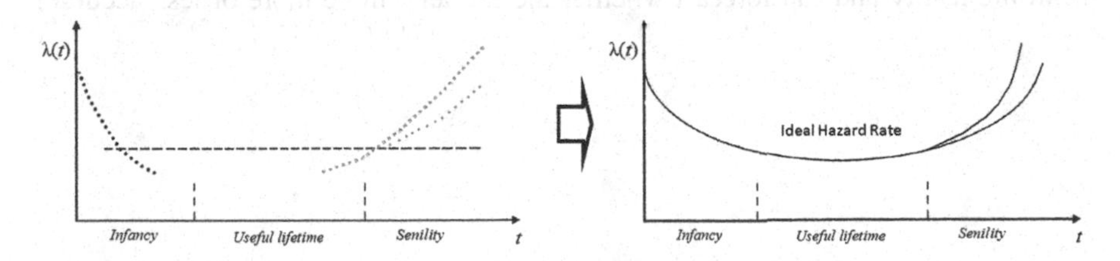

Fig. 8.1 The ideal bathtub as a combination of four functions

8.3.1 The former express common sense understanding of system behavior, and does not comply with a precise logic. By contrast, the latter is established on the basis of theorems and lemmas which infer the conclusions from the premises. Paragraph 1.1.14 offers a fine methodological comparison term; it derives the law of ideal gas from the equations of Boyle, Charles and Avogadro; in a symmetric manner this book deduces the ideal hazard rate by joining equations (8.5).

8.3.2 The three parts of (8.5) derive from distinct and precise sets of assumptions and a unifying analytical equation turns out to be somewhat nonsensical. The application of the tripartite curve is to be related to each group of hypotheses that are typical of infancy, maturity and senescence. Section 6.3.6 presents a detailed comment about the separation of the results.

8.3.3 Scheme (8.1) conflicts with the present theory, as long as we have demonstrated that progressive deterioration influences any phase of the system lifetime. Two sentences in (8.1) are false:

– '*It will generally tend to remain trouble-free over the maturity lifetime*' instead also the activities made by *S* during maturity impair quality and leads to failure;
– '*During ageing components begin to wear out*', broad literature shares this idea, instead ageing does not 'begin to wear' but is affected by the aggravation of wearing due to the cascade effects.

8.3.4 Model (8.5) is based on the assumptions of Lemmas 4.1, 5.1, 6.1, 6.2, and on a.01, a.02 and a.03 that are special additional constraints. Therefore *the present theory does not entail that the tripartite curve is standard* as someone takes for understood; instead this deductive theory yields the following conditions:

1. *The U-shaped curve is true provided that all the hypotheses are true.*
2. *If the hypotheses are partially true, then the U-shaped curve is partially available.*

8.3.5 Let us look into case **2** that is a special quality of the deductive logic.

Researchers deduce the ideal model $\mathcal{X}$ from precise assumptions and those assumptions govern the use of $\mathcal{X}$, an expert addresses an issue with ease when $\mathcal{X}$ perfectly fits with the physical event $\boldsymbol{X}$ (Chap. 1). However, experts are capable of taking advantage of the ideal model $\mathcal{X}$ even when the assumptions do not perfectly cohere with the physical reality [3]. This extended use of $\mathcal{X}$ is allowed by the precision of hypotheses, since one can measure how much the hypotheses diverge from the reality and can forecast whether the model will be more or less accurate.

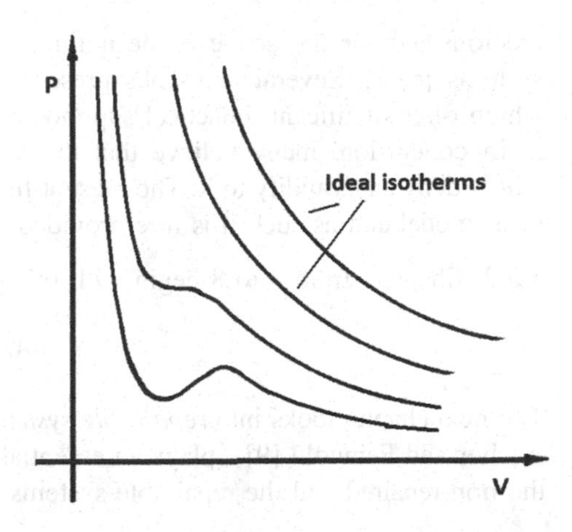

Fig. 8.2 Isotherms of a real gas

Example This is the law of ideal gas that have been demonstrated in Sect. 1.1.4

$$PV = nRT. \tag{1.1}$$

At high temperature each isotherm of a real gas has the form of a hyperbola on a pressure/volume graph. However at low temperature the potential energy due to intermolecular forces becomes more significant compared with the particles' kinetic energy, and the size of the molecules becomes rather significant compared to the empty space between them. The hypotheses of (1.1) progressively diverge from physical reality when the gas temperature drops, and the isotherms exhibit significant deviations from the ideal behavior (Fig. 8.2), yet the extreme values of P and V comply with (1.1) at low temperature and the ideal model may be used with some approximation.

Example This is the basic assumption of the first part of the book

$$S = A_f. \tag{2.7}$$

That it to say, we are looking into the *functioning state* A_f while the recovery state A_r has been ignored. Assumption (2.7) implies that the bathtub curve is the ideal model of *non-repaired systems*, but when one investigates the overall system lifespan, he finds some repair processes. For example, men and women fall sick and cure themselves, however these medical cures do not disprove the Gompertz law which still regulates human ageing.

8.3.6 This just to understand that many real systems come close to the assumptions of (8.5) in a way, and the tripartite curve may be approximated in a large number of works. In modern literature one finds out theorists who often use the U-shaped curve for the purpose of forecasting the performances of systems and addressing issues of dependability. Practitioners adopt it to gain a competitive advantage in the market [4], to improve industrial strategies [5] and so forth. The curve (8.6) offers precious aid for the scope of designing, analyzing, improving and maintaining systems [6, 7]. Several textbooks present the three parts of the system lifetime which offer significant didactical support to teachers and students [8].

In conclusion, many believe that the bathtub curve is a standard model and others deny any validity to it. The present frame demonstrates that the bathtub is an ideal model and as such it is true provided the hypotheses are true.

8.3.7 Chapters from 3 to 8 begin with this physical model

$$S = A_f. \tag{2.7}$$

The next chapter looks into *repairable systems* and this separation—emphasized by Ascher and Feingold [9]—plays an essential role: two distinct ideal models typify the non-repaired and the repairable systems.

References

1. Gnedenko, B. V., Bélyaev, Y. K., & Solovyev, A. D. (1966). *Математические методы в теории надёжности*. Nauka; Translated as: *Mathematical methods in reliability theory*. Academic Press (1969).
2. Nash, F. R. (2016). *Reliability assessments: Concepts, models, and case studies*. CRC Press.
3. Coniglione, F., & Poli, R. (2004). *La realtà modellata. L'approccio idealizzazionale e le sue applicazioni nelle scienze umane*. Franco Angeli.
4. Levin, M. A., & Kalal, T. T. (2003). *Improving product reliability: Strategies and implementation*. Wiley.
5. Smith, D. J. (2007). *Reliability, maintainability and risk: Practical methods for engineers* (7th ed.). Elsevier.
6. Wasserman, G. (2003). *Reliability verification, testing, and analysis in engineering design*. Marcel Dekker Inc.
7. Silverman, M. (2010). *How reliable is your product? 50 ways to improve product reliability*. Super Star Press.
8. O'Connor, P. D. T., & Kleyner, A. (2002). *Practical reliability engineering* (4th ed.). Wiley.
9. Ascher, H., & Feingold, H. (1984). *Repairable Systems Reliability*. Marcel Dekker.

Part II
Reparable Systems

Chapter 9
Properties of Repairable Systems

The previous pages focus uniquely on the single functioning state A_f which runs until the first failure and overlook the state assumed by S when A_f stops.

9.1 The Functioning State and the Recovery State

Here we suppose that S switches from A_f to A_r and vice versa. We are going to investigate a different *physical model* which is usually called the *repairable system*

$$S = (A_f \, OR \, A_r). \tag{2.8}$$

The system S either runs or is under reparation, and thus the probabilities of the states A_f and A_r verify

$$(P_f + P_r) = 1. \tag{9.1}$$

9.1.1 Practitioners normally stop a device when they repair it. Notably the logical connective is the *Exclusive OR* in (2.8), and the states A_f and A_r of artificial systems comply perfectly with it.

Biological systems have a somewhat different conduct. They do not cease living when they are sick and wait to be cured; rather they suspend or slow down the usual activities. A person under medical treatment normally comes to rest, the metabolism slows, and some biological functions become weaker or are even set aside. The A_f state does not completely leave space for A_r but becomes feebler. Practical evidence from the biological domain substantiates (2.8) where the logical connective is the *Inclusive OR* instead of the exclusive *OR*. The latter implies that either one or the other term is true, and a third way is unachievable. The inclusive *OR* means that there is a certain compatibility between the two terms of the statement. Assumption

P. Rocchi, *Reliability is a New Science*, DOI 10.1007/978-3-319-57472-1_9

(2.8) is therefore true for artificial and natural systems alike, albeit with slightly different meanings.

9.2 The Reparability Function

We mean to look into the performances of S which depend on repair processes, hence we calculate the reliability entropy in function of the recovery entropy.

Theorem 9.1 *The reliability and the recovery entropies are mapped by this continuous function*

$$H_f = f(H_r) = \ln(1 - \exp(H_r)) \tag{9.2}$$

We call this *reparability function.*(Fig. 9.1)
Proof Let us obtain P_r from the recovery entropy

$$P_r = \exp(H_r) \tag{9.3}$$

We get probability P_f from (9.1) and combine it with (9.3)

$$P_f = 1 - P_r = 1 - \exp(H_r)$$

We place this equation into the definition of the Boltzmann-like entropy (2.26) and the theorem is proved

$$H_f = H(P_f) = \ln(P_f) = \ln(1 - \exp(H_r))$$

9.2.1 The reparability function has three mathematical properties which will be examined in the order:

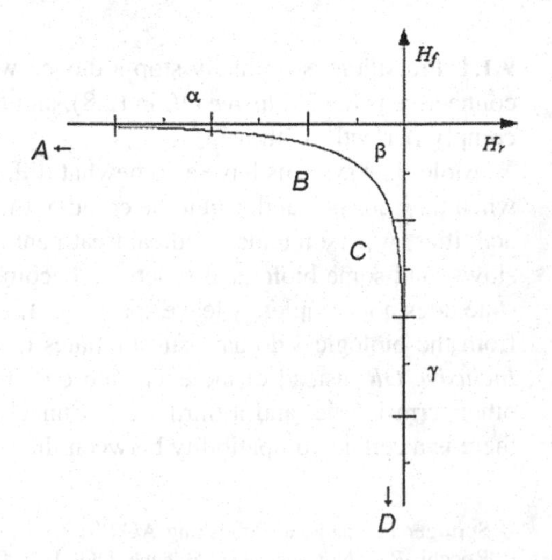

Fig. 9.1 The reparability function

9.3—Property of the Extreme States;
9.5—Property of the Repair Outcomes;
9.7—Property of the Repair Process.

9.2.2 Living systems are self-producing mechanisms which maintain their particular form despite material inflow and outflow, through self-regulation and self-reference. A bacterial cell or a large multicellular organism is able to maintain and renew itself by regulating its composition and conserving its boundaries. The Chilean scientists Humberto Maturana and Francisco Varela introduced the concept of *autopoiesis*—typical of living beings—in opposition to *allopoiesis* which characterizes artificial systems [1]. The allopoietic system brings forth something other than the system itself, whereas the autopoietic system operates on itself.

The reliability entropy H_f qualifies the aptitude of S to work without failures, and the recovery entropy H_r illustrates the disposition of S toward reparation (see 2.29). In substance, the entropies qualify the states of S and not the agents that change the states of S. As a consequence, the reparability function can measure improvement due to the external renewal mechanisms (typical of allopoietic systems), or the benefits of inner agents (typical of autopoietic systems), or both.

9.3 Property of the Extreme States

Denote

$$-\delta = H_r < 0$$

Consider the cases (i) $\delta \to \infty$, (ii) $\delta \to 0$ using Taylor expansions. In case (i) we have

$$H_f = \ln(1 - \exp(-\delta)) \approx -\exp(-\delta) \to 0, \quad \delta \to \infty. \tag{9.4}$$

In case (ii) we have

$$H_f = \ln(1 - \exp(-\delta)) \approx \ln \delta \to -\infty, \quad \delta \to 0. \tag{9.5}$$

Limits (9.4) and (9.5) yield the asymptotes

$$H_r = 0,$$
$$H_f = 0.$$

The asymptotes identify the extreme states of the reparable system (Fig. 9.1)

$$A = (-\infty, 0),$$
$$D = (0, -\infty). \tag{9.6}$$

9.4 Comments on the Extreme States of the System

Let us examine the physical meaning of (9.6).

9.4.1 Statements (2.29) hold:

The reliability entropy expresses the aptitude of S to work without failures.
The recovery entropy illustrates the disposition of S toward reparation.

In substance, two features which are independent in point of logic—the *effectiveness* and the *reparability* of a system—are related by means of the reparability function.

9.4.2 Expression (9.6) tells us that A is the state of a system that functions perfectly when the difficulties in repairing it are null; D is the state of a system incapable of running when the efforts to recover it are immense. The properties of A and D have been derived from (9.1); that is to say the reparability of the system depends on the effective features of S and not on the system age. In principle S can become irreparable when young, or can be easily repaired when old. Although the characteristics of A and D are independent of the system age, writers adopt an heuristic criterion and associate the features of D with a *new system* and the features of D with an *old system*. We share this criterion and conclude:

(I) The new and perfectly functioning system is easily repaired and lies in A.
(II) The old and nonfunctional system is absolutely irreparable and lies in D.

Empirical data substantiates statements (I) and (II), normally new machineries are easily repaired and old ones need many efforts for their repair/maintenance. Physicians recognize the good response of young individuals to cures, which by contrast are far less effective for old patients [2, 3].

Example Let us look into the survival rates of elderly and younger patients who underwent esophageal resection [4]. The first column in Table 9.1 plots the post-operative rates, which are the percentage of all patients who did not die within 60 days of the operation. The second, third and fourth columns display the survival rates registered after 1, 3 and 5 years. The values of the older group are constantly lower than the values of the younger patient group.

9.4.3 In general, S moves from A to D through various stages. The system begins in the state A and declines in quality toward the state D_1 due to the *degeneration principle*. The system is repaired between t_1 and t_2, the functioning entropy

Table 9.1 Survival rates depending on ages. Reproduced from [4] with permission from Elsevier

	Years			
Age	0	1	3	5
< 70	94.6%	71.5%	43.2%	35.3%
≥ 70	89.1%	56.2%	40.7%	32.9%

increases and it restarts in A_1. The entropy H_f diminishes anew until D_2 and after the second repair the system reaches the state A_2. The cycles repeat until the system stops in D where it is considered irreparable. The system lifetime covers the interval $(0, t_k)$.

The scheme includes a number of *failure-repair cycles*; the intervals $(0, t_1)$, (t_2, t_3) etc. refer to the notion of mean time to repair (MTTR), while (t_1, t_2), (t_3, t_4) etc. refer to mean time between failures (MTBF) (Fig. 9.2). Many studies have analyzed the structure of those cycles; researchers drive inquiries about how to optimize system availability, the maintenance methods, testing strategies and so forth [5, 6].

9.4.4 Failure-repair cycles are amply treated in the literature and the previous scheme can be assumed as the *ideal model* of the reparable system's conduct (read Sect. 1.1.4)

The saw–shaped curve is the ideal model for a repairable system's functioning. (9.7)

We note that the U-shaped curve is the ideal model for the functioning system examined in Part 1; and the *saw-shaped curve* is the ideal model of the repairable system analyzed in the second part of the book. These distinct ideal schemes are in accord with the authors who emphasize the different features of systems-with-repair and systems-without-repair such Ascher and Feingold [7].

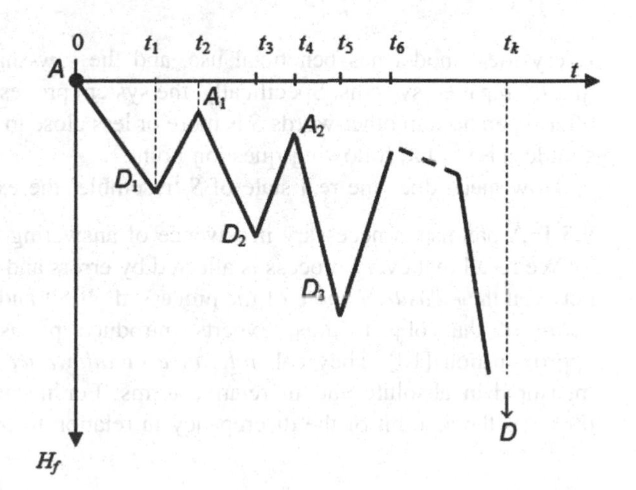

Fig. 9.2 The saw-shaped ideal model of repairable systems

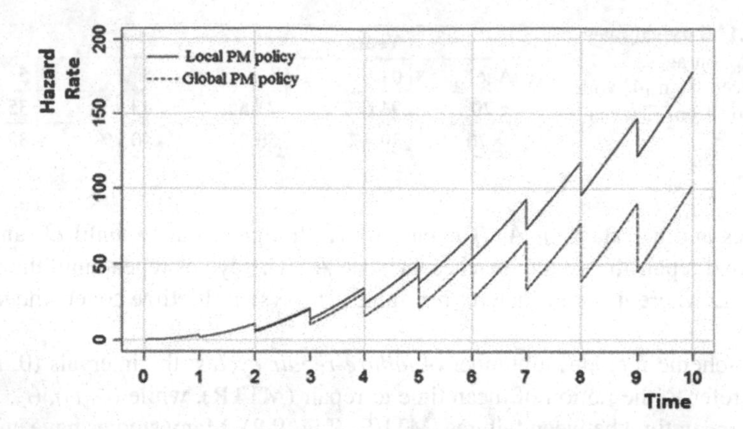

Fig. 9.3 The hazard rates of systems with local PM policy and global PM policy. Reproduced from [9] with permission from springer

The verbal statement (9.7) may be found in several studies which use various parameters in the place of H_f.

Example The failure rate of an electrical appliance exhibits a saw-tooth shape which demonstrates how the reliability improves after maintenance has been performed then the failure rate starts to rise again [8].

Example The case in Fig. 9.3 illustrates the increasing zig-zag hazard rate as a consequence of periodic preventive maintenance (PM) conducted with local and global criteria.

9.5 Property of the Repair Outcomes

Every ideal model has beneficial use, and the saw-shaped curve aids experts to qualify repaired systems. Specifically, the system proves to be more or less efficient after reparation, in other words *S* is more or less close to the model A (or D) and we should answer the following question point:

How much does the real state of *S* 'resemble' the extreme A (or D)?

9.5.1 A premise is necessary in advance of answering this query.

We recall that every process is affected by errors and there is a certain difference between the *measured value* of the processed object and the *theoretical* or *nominal value* of that object; thus, experts introduce precise criteria to qualify this approximation [10]. They call *tolerance* or *allowance* the difference that can be measured in absolute and in relative terms. For instance, the *relative tolerance* provides the amount of the discrepancy in relation to the correct value

$$Relative\ Tolerance = \frac{(Measured\ Value - Theoretical\ Value)}{Theoretical\ Value}$$
$$= \frac{Absolute\ Tolerance}{Theoretical\ Value}.$$

9.5.2 We assume a symmetric criterion to assess the similarity of the system S to A (or D) after the reparation process. We define the *relative deviation* between the repaired system S and the ideal model A (or D) in this manner

$$Relative\ Deviation = \frac{Absolute\ Deviation}{Theoretical\ Value}.$$

Thus we have

$$Absolute\ Deviation = (Theoretical\ Value)(Relative\ Deviation).$$

In engineering projects, ten percent of error or inaccuracy is often accepted as being within engineering tolerance [11]. We fix 10% as the maximum allowed relative deviation—in accordance with the previous pragmatic criterion—and obtain the *maximum absolute deviation* allowed between the repaired system and the ideal model

$$Maximum\ Deviation = (Theoretical\ Value)/10. \tag{9.8}$$

We use the criterion (9.8) to answer the initial query:

How much does the real state of S 'resemble' the extreme A (or D)?

The asymptote $H_f = 0$ is the tangent of the extreme A and forms a straight angle with the axe H_r. We use the straight angle to determine the *theoretical value* of A and obtain the *maximum deviation angle* φ for the tangent of a point that can be associated with the extreme A

$$\varphi = \frac{-180°}{10} = -18°. \tag{9.9}$$

Let us call B the point whose tangent makes the angle φ with the axis H_r (Fig. 9.3). In a similar way we fix the point C whose tangent makes the *maximum deviation angle* ψ with the axis H_f

$$\psi = 18° \tag{9.10}$$

Let us calculate the derivatives of function (9.2)

$$\frac{dH_f}{dH_r} = \frac{-\exp(H_r)}{(1 - \exp(H_r))} < 0.$$
$$\frac{d^2H_f}{dH_r^2} = \frac{-\exp(H_r)}{(1 - \exp(H_r))^2} < 0, \qquad H_r < 0.$$

The derivatives prove that the reparability function decrease continuously and monotonously from 0 to $-\infty$ when the variable steps up from $H_r = -\infty$ to $H_r = 0$. So for any $h < 0$ there is single $s = s(h)$ so that

$$\frac{dH_f(s)}{dH_r} = h,$$

$$s(h) = \ln\left(\frac{-h}{1-h}\right).$$

Consequently we may choose $h_\alpha < 0$, $h_\gamma < 0$ so that $|h_\alpha|$ is sufficiently small and $|h_\gamma|$ is sufficiently large and determine the zones α and γ (Fig. 9.1)

$$\alpha = \{H_r : H_r \in (-\infty, s(h_\alpha))\}, \quad \gamma = \{H_r : H_r \in (s(h_\gamma), 0)\}.$$

In the zone α which characterizes asymptote $H_f = 0$ we have the inequality

$$h_\alpha < \frac{dH_f(H_r)}{dH_r} < 0.$$

In the zone γ which characterizes asymptote $H_r = 0$ we have the inequality

$$-\infty < \frac{dH_f(H_r)}{dH_r} < h_\gamma.$$

Let us assume α as the interval (*A*, *B*), and γ as (*C*, *D*). Using the values (9.9) and (9.10) we get h_α and h_γ

$$\begin{aligned} \varphi &= -18^o, \\ h_\alpha &= tg\,\varphi = -0.3249, \\ (-90^o + \psi) &= -72^o, \\ h_\gamma &= tg(-72^o) = -3.0777. \end{aligned} \tag{9.11}$$

Then we have that coordinates of points *B* and *C* are the following

$$\begin{aligned} H_r(B) &= -1.4055, \quad H_f(B) = -0.2813, \\ H_r(C) &= -0.2813, \quad H_f(C) = -1.4054. \end{aligned} \tag{9.12}$$

On the basis of statements (I) and (II) the following conclusions can be drawn (Fig. 9.4):

The system which lies in α approximates the ideal case A and is as–good–as–new (AGAN).
The system which lies in γ approximates the ideal case D and is as–bad–as–old (ABAO).

(9.13)

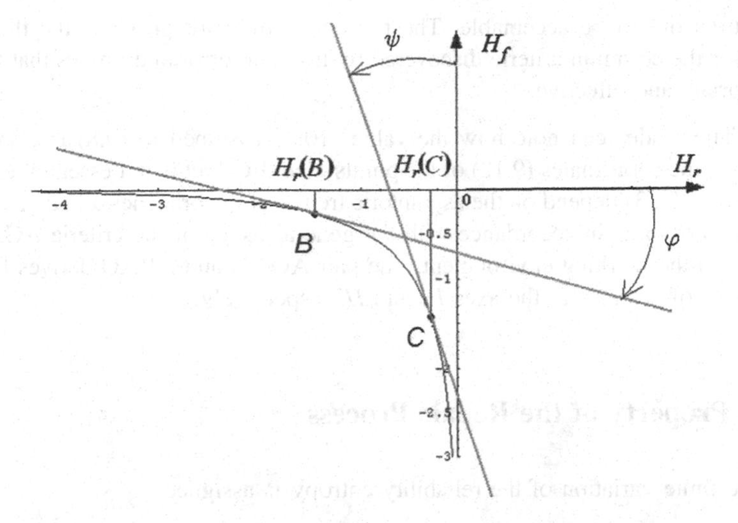

Fig. 9.4 Method to determine the points *B* and *C* Reproduced from [12] with permission from Scientific Research Publishing, ©SciRe

9.6 Comments on the Repair Outcomes

Definitions (9.13) fit with coeval authors who adopt the criteria AGAN and ABAO on the basis of professional practice. Experts associate the state 'as-good-as-new' with a system perfectly repaired (or maintained). The interventions for *S* usually are very expensive, but result in an efficient outcome. On the other side, the 'as-bad-as-old' system is often judged as the consequence of an imperfect repair which has the advantage of being cheap or fast. Telling economic and organizational criteria sustain the notions of AGAN and ABAO in the working environment [13] and turn out to be perfectly consistent with the present theoretical frame.

9.6.1 The reparability function $H_f = f(H_r)$ is continuous in the domain $(0, -\infty)$ and demonstrates that in abstract a system can assume infinite degrees of qualities after reparation. A few authors share this view such as Malik [14] who includes intermediate degrees of quality between AGAN and ABAO.

9.6.2 The majority of writers overlooks intermediate results between AGAN and ABAO [15, 16]. Someone has fixed the so-called *(p, q) rule*, which formalizes how AGAN and ABAO are mutually exclusive. This rule holds that when *S* becomes AGAN with probability *p*, then it becomes ABAO with probability $q = (1 - p)$ [17]. The curve of the reparability function shows visually how this second approach is correct under a good approximation criterion. In fact, the horizontal zone α and the vertical zone γ of the reparability function are separated by the intermediate range β which (9.12) proves to be rather small in size. Assuming the distribution of the physical systems is uniform along the curve $H_f = f(H_r)$, the systems placed in β should be judged as a minority group and the *(p, q)* *rule* turns out to be acceptable. The present framework provides the theoretical basis for the common criteria discovered by trial and error and proves that they are appropriate and effective.

9.6.3 The reader can note how the value '10%' assumed in (9.8) is essential to calculate the coordinates (9.12) of the points *B* and *C*, but is not essential to (9.13). Properties (9.13) depend on the asymptotic trends of the branches α and γ, and have general meanings in accordance with the general usage of the criteria AGAN and ABAO in the working environment. The pair AGAN and ABAO derives from the closeness of α and γ to the axes H_r and H_f respectively.

9.7 Property of the Repair Process

Let the finite variation of the reliability entropy is assigned

$$\Delta H_r = -k, \quad k > 0 .$$

And introduce the finite difference quotient

$$\Delta H_f(x) = \frac{H_f(x-k) - H_f(x)}{(-k)}, \quad x < -k.$$

This finite quotient has the following property

$$\frac{dH_f(x)}{dH_r} \leq \Delta H_f(x) \leq 0. \tag{9.14}$$

Consequently one obtains

$$\Delta H_f(x) \rightarrow 0, \quad x \rightarrow -\infty. \tag{9.15}$$

$$\Delta H_f(x) \rightarrow -\infty, \quad x \rightarrow -k. \tag{9.16}$$

Equations (9.15) and (9.16) mean in the order that

$$\begin{aligned} &|\Delta H_f(x)| \text{ is 'very small'} \quad \text{if } x \rightarrow -\infty, \\ &|\Delta H_f(x)| \text{ is 'very large'} \quad \text{if } x \rightarrow -k. \end{aligned} \tag{9.17}$$

9.8 Comments on the Repair Process

Let us examine the practical impact of the last results.

9.8.1 Equation (9.17) holds that the amelioration of appliances obtained after an identical repair process depends on the initial state of those appliances. Two identical interventions are able to bring very different benefits. We demonstrate this statement using the following concrete example case.

Example Workers replace a precise component with a new one in a group of appliances of the same kind. Suppose the effort required by this repair is indifferent to the status of appliances and is constant

$$|\Delta H_r| = k, \quad k > 0.$$

When the appliance is placed in α, (9.15) demonstrates that the change ΔH_f caused by ΔH_r is insignificant (see segment *m* in Fig. 9.5). Actually the new component results in a little amelioration since it substitutes a part of the AGAN system.

Now let the physical system be placed in γ, Eq. (9.16) demonstrates that the repair *k* results in conspicuous improvements (see segment *M* in Fig. 9.4). In fact workers replace a very old component of *S* with a new one.

9.8.2 Normally a system which needs to be repaired, lays in the range γ. Suppose *S* is placed in *X* (Fig. 9.6), one can infer that the generic change ΔH_r—caused by the restoration—can prompt the following typology of results, depending on the end-state of *S*:

1. The repaired system moves into the range *A–B* It works better after the intervention; and becomes as-good-as-new. This result means that the repair or maintenance is *perfect*.
2. The system migrates in the range *B–C* and thus has intermediate quality. A treatment restores the system operating state to be somewhere between as-good-as-new and as-bad-as-old. This result means that the repair or maintenance is *imperfect*.

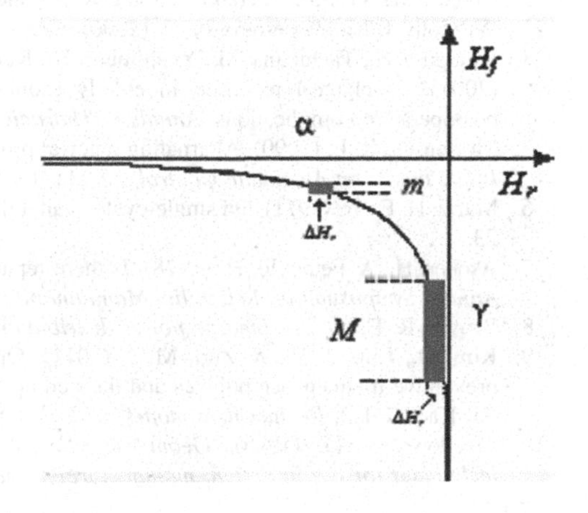

Fig. 9.5 Different outcomes obtained after two identical repairs. Reproduced from [12] with permission from Scientific Research Publishing, ©SciRe

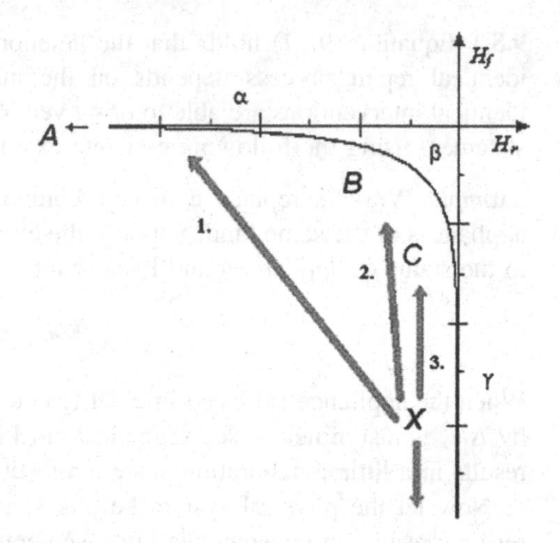

Fig. 9.6 Potential outcomes after a repair process. Reproduced from [12] with permission from Scientific Research Publishing, ©SciRe

3. The system remains in the zone γ. This means that *S* is still bad-as-old and the repair is *minimal*.
4. The system falls below *X*. The system's operating conditions become worse than the conditions observed prior to the restoration/maintenance operations. The repair process has damaged the system status.

The reparability function provides results from 1 to 4 which exhaust the forms of repairs in agreement with the current literature [18].

References

1. Maturana, H. R., & Varela, F. J. (1980). *Autopoiesis and cognition: The realization of living*. Dordrecht: Reidel.
2. Cole, W. H. (1953). Operability in the young and aged. *Annals of Surgery, 138*(2), 145–157.
3. Wilmoth, J. M., & Ferraro, K. F. (2006). *Gerontology: Perspectives and issues*. Springer.
4. Kinugasa, S., Tachibana, M., Yoshimura, H., Kumar Dhar, D., Shibakita, M., Ohno, S., et al. (2001). Esophageal resection in elderly esophageal carcinoma patients: improvement in postoperative complications. *Annals of Thoracic Surgery, 71*, 414–418.
5. Mortensen, R. E. (1990). Alternating renewal process models for electric power system loads. *IEEE Trans. on Automatic Control, 35*(11), 1245–1249.
6. Martz, H. F., Jr. (1971). On single-cycle availability. *IEEE Trans. on Reliability, R-20*(1), 21–23.
7. Ascher, H., & Feingold, H. (1978). Is there repair after failure? In *Proceedings. of the IEEE Annual Symposium on Reliability Maintainability* (pp. 190–197).
8. Brown, R. E. (2008). *Electric power distribution reliability*. CRC Press.
9. Kim, D., Lim, J. H., & Zuo, M. J. (2011). Optimal schedules of two periodic imperfect preventive maintenance policies and their comparison. In L. Tadj, M.S. Ouali, S. Yacout, D. Ait-Kadi (Eds.), *Replacement models with minimal repair* (pp. 141–161). Springer.
10. Meadows, J. D. (1995). *Geometric dimensioning and tolerancing: Applications and techniques for use in design, manufacturing, and inspection*. Marcel Dekker.

11. A. A. V. V. (1984). *Military handbook: evaluation of contractor's calibration system*. U.S.: Department of Defense.
12. Rocchi, P., & Tsitsiashvili, G. (2012). Four common properties of repairable systems calculated with the Boltzmann-like entropy. *Applied Mathematics, Special issue on Probability and Its Applications, 3*(12A), 2026–2031.
13. Sim, S. H., & Endrenyi, J. (1993). A failure-repair model with minimal and major maintenance. *IEEE Transactions on Reliability, 42*(1), 134–140.
14. Malik, M. A. K. (1979). Reliable preventive maintenance policy. *AIIE Transactions, 11*(3), 221–228.
15. Wang, H. (2002). A survey of maintenance policies of deteriorating systems. *European Journal of Operational Research, 139*(3), 469–489.
16. Bloch, H. P. (1988). *Improving machinery reliability* (2nd ed.). Gulf Professional Publishing.
17. Chan, P. K. W., & Downs, T. (1978). Two criteria for preventive maintenance. *IEEE Transactions on Reliability, R-27*, 272–273.
18. Brown, M., & Proschan, F. (1983). Imperfect repair. *Journal of Applied Probability, 20*, 851–859.

Part III
Overall Conclusion

Chapter 10
Conclusive Remarks and Comments

The reliability theory, despite the enormous amount of work done, has not yet reached the status of science. It likens a basket of useful tools and effective mathematical outcomes, which however do not find order inside a Cartesian frame.

10.1 What We Meant to Do

Gnedenko, Soloviev and Bélyaev start with the Markov model and establish that the system *reliability* is the general exponential function

$$P_f(t) = e^{-\int_0^t \lambda(t)dt} \tag{10.1}$$

They obtain (10.1) following the deductive process of this kind

$$\textit{Chained operations} \Rightarrow P_f(t) \textit{ is general exponential function}; \lambda(t) \textit{ is generic.} \tag{10.2}$$

This achievement begins to order the scientific concepts into their proper places and can be defined as the *first law of reliability*.

Gnedenko and his colleagues establish the various trends of $\lambda(t)$ in generic terms and the present book has the purpose of completing the deductive construction initiated by the eminent authors of the Russian school.

10.1.1 This work begins by resuming the principal traits of a theory based on principles or laws. It centers on *logical implications* that relate the consequences to the causes with precision. In addition a principle-based theory has the following typical features:

P. Rocchi, *Reliability is a New Science*, DOI 10.1007/978-3-319-57472-1_10

- It begins with *physical models* that depict the objects under examination;
- It fixes the properties of the *physical models* using formal *principles*;
- It defines *ideal models*.

10.1.2 This book introduces an original armory of formal tools which includes:

- *The structure of levels,*
- *The Boltzmann-like entropy,*

The first model allows us to zoom-in and zoom-out on a dynamic system. One can overview the system as a whole and can dissect its smallest elements. The second function stems from the well-known Boltzmann entropy which is used to illustrate the devolution of the thermodynamic system.

10.2 What Principles

This work poses the following basic principles:

- *The Principle of Superposition.*
- *The Principle of Progressive Deterioration* $\rightarrow$ *Linearity Assumption.*
- *The Principle of Random Accidents* $\rightarrow$ *Binomial Assumption.*

The first offers the 'inclusive view' of system dependability for consideration as long as a broad variety of failure factors overlap without any order and bring *S* to its end. The second principle holds that collateral effects harm the operations of *S* with continuity, and the *linearity assumption* qualifies this rule. The third principle describes random agents that lead the system to cease functioning.

10.3 What We Have Found in Part 1

Lemmas 4.1, 5.1, 6.1 and 6.2 demonstrate that the hazard rate follows three principal courses depending on four precise premises

$$\begin{aligned}&\textit{Linearity Assumption or Binomial Assumption} \Rightarrow \lambda(t) \textit{ is constant.}\\&\textit{Linearity Assumption and Linear Cascade Effect} \Rightarrow \lambda(t) \textit{ is power function.}\\&\textit{Linearity Assumption and Composite Cascade Effect} \Rightarrow \lambda(t) \textit{ is exponential function.}\end{aligned} \tag{10.3}$$

These results are consistent with the general statement (10.1) and provide details about it.

10.3.1 The hazard rates functions listed in (10.3) derive from rigorous hypotheses, and are able to assess the tripartite curve as the ideal model of A_f

$$\textit{The U-shaped curve is the ideal model of system's hazard rate.} \qquad (8.6)$$

We do not unify the three hazard rates (10.3) in a single function because each one derives from distinct assumptions that are to be verified singly.

10.3.2 The present frame presents the bathtub curve as *an ideal model* and *not as a standard rule*, that is to say it is true on the condition that the assumptions are true. As matter of fact, a few systems comply perfectly with the hypotheses of the bathtub curve; but some systems approximate those hypotheses and the bathtub curve makes the study of the system dependability easier. Several treatises, essays and articles show how the tripartite curve (8.6) offers benefits to experts who look into natural and artificial systems, or design a machine or arrange a process in view of the objectives they want to achieve.

10.3.3 The conclusions of the first part of the book are considerably close to the following studies:

- Thermodynamic reliability (Sect. 2.4.4),
- Network theory of aging (Sect. 3.6.1),
- Degeneration theory (Sect. 4.1),
- Degradation reliability methods (Sect. 4.2),
- Physics-of-failure methods (Sect. 4.4.2),
- Competing causes of failure (Sect. 5.5.3),
- Cumulative and extreme shock models (Sect. 5.6),
- The theory of aging and longevity (Sect. 6.5.7).

10.3.4 The present work does not create a new theory or founds a new science; it merely achieves the objectives declared in Chap. 1. The calculations of the different trends of $\lambda(t)$ conform to the deductive approach of Boris Gnedenko, Alexander Soloviev and Yuri Bélyaev and contribute to completing what they produced.

10.4 What We Have Found in Part 2

The reliability entropy is mapped with the recovery entropy under the assumption that S switches between two alternating states

$$\textit{Functioning and repair states are alternating} \Rightarrow \textit{The reparability function}. \qquad (10.4)$$

10.4.1 Using the reparability function we have defined the behavior of systems that cross various failure-repair cycles

The saw-shaped curve is the ideal model of repairable system's behavior. (9.7)

10.4.2 The reparability function proves the following notions on the theoretical plane:

- Two distinct features of systems—the *performances* and the *reparability*—are inversely proportional.
- The outcomes of repair processes approximate two opposite levels of quality, called *as-good-as-new* and *as-bad-as-old.*
- An identical repair intervention brings forth significant improvement of an ABAO system, and negligible amelioration of an AGAN system.
- A repair process is able to bring forth a *perfect, imperfect, minimal* or a *worse result.*

The scientific community has discovered these concepts by trial and error and even by intuition. The reparability function brings added value since heuristic criteria provide somewhat unrelated concepts that need order and a unifying scheme.

10.4.3 The conclusions drawn in the present part of the book are considerably close to the researches on:

- Management of repair methods,
- Availability calculations,
- Reliability centered maintenance.

10.4.4 The results obtained in the second part of this book make the flexibility of the Boltzmann-like entropy clear.

10.5 Overall Discussion of the Results

European schools have always taken into high consideration the philosophical meanings of the physical laws established by theorists and a non-negligible number of scientists emerge as trustworthy thinkers. For instance, Ludwig Boltzmann (1844–1906) was appointed as the reader of the *Science Philosophy* course, in addition to his teaching in Mathematical Physics. More recently, Carl Friedrich von Weizsäcker (1912–2007), a nuclear physicist, was professor of *Philosophy* at the University of Amburg from 1957 to 1970. One can even recall how a characteristic of mathematicians in Moscow was their interest in philosophy during the 19th century. Literature remembers Nikolaj Bugaev (1837–1903) "as much a philosopher as a mathematician". The first course in the theory of functions of real variables at Moscow University was held in 1900 and a significant role in creating a climate of interest in this theory was played by the philosopher, engineer and mathematician Pavel Florensky (1882–1937).

10.5.1. *What has this got to do with philosophy?*
Despite the apparent differences emerging between philosophy and science, they go together in search of knowledge; they try to explain what there is in the world and sometimes proceed side by side. Popper claims that empirical research is impossible without general ideas and adds that ideas which initially 'floated' in the regions of metaphysics, were transformed into important scientific doctrines:

How can this happen?

Technicians work with concrete things, while scientists—among other goals—aim at widening the horizon of understanding. When scientists broaden the view of the world, then a sort of spontaneous cooperation occurs with philosophers who are attracted by general matters.

Philosophy also proposes methodological criteria to improve scientific knowledge and erudition. As an example, we quote the *Methods* of Mill [1] which are intended to illuminate the issues of causation. Hence, when a scientific argument appears to be insoluble, there are those who try to approach it philosophically. See, for instance, the current discussion of the principles of quantum mechanics addressed by thinkers and mathematicians.

Broadly speaking, there are two sorts of questions concerning every scientific discipline [2]. The first sort are the *first-order questions* which regard the subject matter of the discipline; the *second-order questions* concern the methodology, knowledge, history, social impact and philosophy of the discipline. These arguments are usually addressed by thinkers and humanists who often argue on the intricate epistemological query: "how do we know what we know?" This kind of topics are alien to mathematics and mathematicians usually overlook them, instead the present work has made attempts to confront some second-order questions, for instance:

- Deductive logic has been illustrated in order to show the method employed to gain knowledge here.
- Inferences (10.2), (10.3) and (10.5) explain how some system behaviors take origin.
- The *deterioration arrow of time* provides insight into the cosmological study of the world and emerges as a typical philosophical argument.
- A short annotation summarizes the probability interpretation issue and a solution in Sect. 2.4.4.

10.5.2 *Positive philosophy and positive science*

The reliability field crosses engineering and biology and the contrasting methodologies that they use, raise some issues in the scientific community.

Engineering and technology comply with the *positivist* thought in a perfect manner [3]. Engineering and technology are based on the assumption that there are objective entities in the physical reality, and that material entities are external to the researcher and therefore knowable in their real essence. The goal of researchers is to get to the formulation of statements based on the categories of cause and effect, and a theory can always be falsified. Engineers also prefer to adopt the mathematical language, which employs abstract and relatively simple objects—e.g. numbers, sets and functions—and these abstract items simplify the reality in a way, but have the virtue of depicting the phenomena with precision.

Actions carried out by organisms are not simply determined by 'laws'. Sometimes animate beings lead to theories that cannot be falsified and often cannot be formulated with an adequate degree of precision. Human sciences—e.g. geriatrics, medicine, and sociology—also accept subjectivist philosophies, including *interpretivism* and *constructivism* [4]. On the other side, they do not reject the positivist approach, since the precision of mathematics is an extraordinary feature capable of controlling present and future events.

The results obtained here conform to the positivist stance. We are conscious that the present mathematical models oversimplify the complex behavior of living beings; in opposition to this defect the mathematical logic has the advantage of offering precise terms of reference to doctors, actuaries, sociologists etc. Mathematical principles help researchers to understand the mechanisms of aging, mortality, survival, and longevity of living beings that are not readily explainable otherwise.

The application of mathematical models—inferred through deductive logic here—to the biological domain is not automatic but require accurate analysis and the present book has developed some accurate comments. For example,

- Section 2.1.7 discusses whether stochastic models are appropriate representations of living beings;
- Section 9.1 compares the functioning and the recovery states which occur alternatively during the system lifetime;
- Section 9.2.2 makes a comment on autopoietic systems as opposed to allopoietic systems;
- Section 6.5.6 distinguishes senescent mortality factors from non-senescent factors.

10.5.3 *Science and knowledge organization*

The Academic Press Dictionary of Science and Technology [5] defines *science* as:

> The systematic observation of natural events and conditions in order to discover facts about them and to formulate laws and principles based on these facts.

This entails that science involves the acquisition of knowledge and also it provides a certain logical order. The present work, congruent with this, lays a pair of 'logical capstones'.

1. As first we recall that the principal causes of failure for systems are *the progressive deterioration of components* and *sudden random attacks.*
2. As second, this book puts forward two ideal models. On the one hand, the *bathtub curve* describes the ideal behavior of systems and circumvents the debate emerging between those who deem this curve as a standard, and those who reject it because of the broad variety of irregular $\lambda(t)$ experienced in the physical reality. On the other hand, the reparability function leads to the *saw-shaped curve* that depicts the basic properties of repaired and maintained systems.

The pair of causes cited in point 1 call to mind the *Pareto Principle.* In the early twenty century, Italian economist Vilfredo Pareto observed that 20% of the Italian population owned 80% of property in Italy. Some decades later the American J.M. Juran generalized Pareto's remark and suggested that a small number of causes determine most of the results in any situation. This rule, named after Pareto, states that the majority of consequences stem from a small number of causes for usual—economic, social, technical, psychological etc.—phenomena. This idea is not so regrettable in the reliability domain since we have seen how two principal root-causes regulate the reliability of systems.

The concise principles and rules illustrated in the present book should enable this domain to evolve from being a 'heap of stones' to a 'house of stones' that has the essential virtue of being organized logically, consistently and effectively.

10.5.4. *Multiplicity and unity*
It may be said that there are two principal courses of study in the reliability domain. Technicians investigate the specific and physical reasons for failures (physics of failure); they follow an *analytical* approach but do not say anything about the broad properties of systems On the other side, theorists look through systems' behavior using statistical methods. They provide *unified* descriptions of systems but do not elucidate the physical mechanism of faults.

The present work has the virtues of both methods. It is *analytical* since it starts from detailed assumptions about the components of S, it proves how each trend of the hazard rate derives from a precise root-cause and bridges the failure prediction with the reliability studies for POF. At the same time *the present frame provides a unified view* of system behaviors and draws general conclusions no matter the systems are artificial or living. E.g. we have demonstrated that wearing and random accidents are the universal causes of constant $\lambda(t)$. E.g. a variety of failure factors overlap with different rules and result in the various forms of the hazard rate function that experimentalists observe in the physical reality.

10.5.5. *Corroboration and collaboration*
The present book means to complete the seminal work of Gnedenko, and must undergo rigorous testing. For Popper, the authentic scientific process begins with

the validation of the logical scheme and emphasizes that a hypothesis can be falsified by a single negative instance [6].

All scientific theories are provisional and the destiny of the present construction goes through accurate screening. Practically, we have begun the test phase by examining a variety of practical cases. If this theory will survive falsification, it will be classified as a contribution suitable for the reliability science.

Presently, empirical induction dominates the province of depedablity. Current literature offers a wealth of results, mostly obtained through data-driven inquiries, and modern statistical inquiries provide a large number of disposable outcomes which can be used to validate the present frame.

References

1. Mackie, J. L. (1967). Mills methods of induction'. In P. Edwards (ed.), *The encyclopedia of philosophy* (Vol. 5, pp. 324–332).
2. Corry, L. (1989). Linearity and reflexivity in the growth of mathematical knowledge. *Science in Context, 3*(2), 409–440.
3. Friedman, M. (1999). *Reconsidering logical positivism*. Cambridge University Press.
4. Schaffner, K. F. (1993). *Discovery and explanation in biology and medicine*. Chicago: University Press.
5. A. A. V. V. (1992). *Academic press dictionary of science & technology*. Academic Press.
6. Corvi, R. (1997). *An introduction to the thought of Karl Popper*. Routledge.

Appendix A
History of Reliability in Literature

This bibliographical appendix includes selected works which cover a period of time from the early beginnings of the reliability theory to present day.

Azarkhail, M., & Modarres, M. (2012). The evolution and history of reliability engineering: Rise of mechanistic reliability modeling. *International Journal of Performability Engineering, 8*(1), 35–47.

Barlow, R. E. (1984). Mathematical theory of reliability: A historical perspective. *IEEE Transactions on Reliability, R-33*(1), 16–20.

Bhamare, S. S., Yadav, O. P., & Rathore, A. (2007). Evolution of reliability engineering discipline over the last six decades: A comprehensive review. *International Journal of Reliability and Safety, 1*(4), 377–410.

Brennan, R. L. (2001). An essay on the history and future of reliability from the perspective of replications. *Journal of Educational Measurement, 38*(4), 295–317.

Demiyanushko, I. (2011). The theory of reliability of machines and mechanism: History, state-of-the-art and prospects. *Proceedings of the 13th Congress on Mechanism and Machine Science*, 1–6

Ebel, G. (1998). Reliability physics in electronics: A historical view. *IEEE Transactions on Reliability, 47*(3), 379–389.

Fukuda, M. (2000). Historical overview and future of optoelectronics reliability for optical fiber communication systems. *Microelectronics Reliability, 40*(1), 27–35.

Gnedenko, E., & Ushakov, I. (2013). B. V. Gnedenko: The father of the Russian school of reliability theory. *Automatic Control and Computer Sciences, 47*(2), 57–61.

Hambleton, R. K., & Slater, S. C. (1997). Reliability of credentialing examinations and the impact of scoring models and standard-setting policies. *Applied Measurement in Education, 10*(1), 19–38.

Juran, J. M. (1995). *A history of managing for quality: The evolution, trends, and future directions of managing for quality*. ASQC Quality Press.

Mclinn, J. (2011). A short history of reliability. *Journal of the Reliability Information, 22*, 8–15.

Ohring, M. (1998). *Reliability and failure of electronic materials and devices*. Academic Press.

Peck, D. S., & Zierdt, C. H. Jr. (1974). The reliability of semiconductor devices in the Bell System. *Proceedings of the IEEE, 62*(2), 185–211.

Rukhin, A. L., & Hsieh, H. K. (1987). Survey of Soviet work in reliability. *Statistical Science, 2* (4), 484–495.

Rueda, A. (2004). Pioneers of the reliability theories of the past 50 years. *Proceedings of the Annual Symposium on Reliability and Maintainability*, 102–109.

Saleh, J. H., Marais, K. (2006). Highlights from the early (and pre-) history of reliability engineering. *Reliability Engineering and System Safety, 91*(2), 249–256.

P. Rocchi, *Reliability is a New Science*, DOI 10.1007/978-3-319-57472-1

Saleha, J. H., Maraisb, K. B., Bakolasa, E., & Cowlagia, R. V. (2010). Highlights from the literature on accident causation and system safety: Review of major ideas, recent contributions, and challenges. *Reliability Engineering & System Safety, 95*(11), 1105–1116.

Ushakov, I. (2000). Reliability: Past, present, future. *Proceedings of the 2nd Conf. Mathematical Models in Reliability*, 3–21.

Ushakov, I. (2007). Is reliability theory still alive?. *Reliability: Theory& Applications, 2*(1), March 2007.

Ushakov, I. (2012). Reliability theory: History & current state in bibliographies. *Reliability: Theory & Applications, 7*(1), March 2012.

Zio, E. (2009). Reliability engineering: Old problems and new challenges. *Reliability Engineering & System Safety, 94*(2), 125–141.

Appendix B
Two Sided Estimates of Rate Convergence in Limit Theorem for Minimums of Random Variables

by Gurami Sh. Tsitsiashvili

B.1. Introduction

In this appendix, Weibull and Gompertz distributions are used as limit distributions in the scheme of minimum of independent and identically distributed random variables (i.i.d.r.v.'s). These distributions are considered as lifetime distributions [1, 2]. The entropy function which gives substantial explanation of their applications in this book does not give convenient tool for estimates of rate convergence in scheme of minimum. So it is worthy to add this approach by technique of rate convergence estimates in limit theorems of probability theory. Importance of such considerations is confirmed by manifold articles and monographs of Gnedenko [3] who paid large attention of probability theory and their applications including reliability theory.

B.2. Estimates of closeness between limit and pre-limit distributions in scheme of random variables minimum

Assume that $U_1,\ldots,U_n,\ldots$ is the sequence of independent and identically distributed random variables (i.i.d.r.v.'s) with common distribution function (d.f.) $F(t) = P(U_1 \leq t)$. Suppose that for some positive a and b we have $\bar{F}(t) = \exp(-at^b)$. (Here and further for nonnegative number f put $\bar{f} = 1 - f$). That means d.f. $F(t)$ is Weibull distribution. Direct calculations give us that following random variables (r.v's) coincide by distribution $(=^d)$

$$n^{1/b}\min(U_1,\ldots,U_n) =^d U_1. \tag{1}$$

Assume that $V_1,\ldots,V_n,\ldots$ is the sequence of nonnegative i.i.d.r.v.'s with common d. f. $G(t) = P(V_1 \leq t), \bar{G} = 1 - G, \bar{G}(0) = 1$. Denote $v = t^b$ and put

P. Rocchi, *Reliability is a New Science*, DOI 10.1007/978-3-319-57472-1

$$\bar{Q}(v) = \bar{G}\left(v^{1/b}\right),$$

$$\bar{G}_n(t) = P\left(n^{1/b}\min(V_1,\ldots,V_n) > t\right),$$

$$\Delta'_n = \sup_{t\geq 0}|\bar{F}(t) - \bar{G}_n(t)|.$$

Our problem is to investigate the convergence to zero of Δ'_n

Theorem B.1 *This equality is true*

$$\Delta'_n = \sup_{v\geq 0}|\exp(-nav) - \bar{Q}^n(v)|.$$

Proof

$$\begin{aligned}
\Delta'_n &= \sup_{t\geq 0}\left|P\left(n^{1/b}\min(X_1,\ldots,X_n) > t\right) - P\left(n^{1/b}\min(Y_1,\ldots,Y_n) > t\right)\right| \\
&= \sup_{t\geq 0}\left|P\left(\min(X_1,\ldots,X_n) > tn^{-1/b}\right) - P\left(\min(Y_1,\ldots,Y_n) > tn^{-1/b}\right)\right| \\
&= \sup_{t\geq 0}|P(\min(X_1,\ldots,X_n) > t) - P(\min(Y_1,\ldots,Y_n) > t)| \\
&= \sup_{t\geq 0}|\bar{F}^n(t) - \bar{G}^n(t)| = \sup_{t\geq 0}|\exp(-nat^b) - \bar{G}^n(t)| \\
&= \sup_{v\geq 0}|\exp(-nav) - \bar{Q}^n(v)|.
\end{aligned}$$

Assume that $X_1,\ldots, X_n,\ldots$ is the sequence of i.i.d.r.v.'s with common d.f. $F(t) = P(X_1 \leq t)$, and for some positive numbers a, b, c

$$\bar{F}(t) = \exp(-a\exp(b(t-c))).$$

This means that d.f. $F(t)$ is Gompertz distribution. Direct calculation show that

$$\min(X_1,\ldots,X_n) + \ln nb =^d X_1. \tag{2}$$

Suppose that $Y_1,\ldots, Y_n,\ldots$ is the sequence of i.i.d.r.v.'s with common d.f. $G(t) = P(Y_1 \leq t)$. Denote

$$v = \exp(b(t-c)).$$

and put

$$\bar{G}_n(t) = P(\min(Y_1, \ldots, Y_n) + \ln nb > t),$$

$$\bar{Q}(v) = \bar{G}(\ln vb + c),$$

$$\Delta''_n = \sup_{-\infty < t < \infty} |\bar{F}(t) - \bar{G}_n(t)|.$$

Theorem B.2 *This equality is true*

$$\Delta''_n = \sup_{v \geq 0} |\exp(-nav) - \bar{Q}^n(v)|.$$

Proof Indeed,

$$\begin{aligned}
\Delta''_n &= \sup_t |P(\min(X_1, \ldots, X_n) + \ln nb > t) - P(\min(Y_1, \ldots, Y_n) + \ln nb > t)| \\
&= \sup_t |P(\min(X_1, \ldots, X_n) > t - \ln nb) - P(\min(Y_1, \ldots, Y_n) > t - \ln nb)| \\
&= \sup_t |P(\min(X_1, \ldots, X_n) > t) - P(\min(Y_1, \ldots, Y_n) > t)| \\
&= \sup_t |\bar{F}^n(t) - \bar{G}^n(t)| \\
&= \sup_t |\exp(-na \exp(b(t-c))) - \bar{G}^n(t)| \\
&= \sup_{v \geq 0} |\exp(-nav) - \bar{Q}^n(v)|.
\end{aligned}$$

Remark B.1 From Theorems 1 and 2 we have the equalities

$$\Delta'_n = \Delta''_n = \sup_{v \geq 0} |\exp(-nav) - \bar{Q}^n(v)|.$$

Which allow to estimate rate convergence to zero for $n \to \infty$ of the expressions Δ'_n, Δ''_n using similar conditions on the function $q(v) = \exp(-av) - \bar{Q}(v)$. Further denote Δ'_n, Δ''_n as Δ_n.

B.3. Upper and Low Bounds for Convergence of Δ_n to Zero

Fix $a \geq 0$ and designate

$$A(v) = av,$$

$$e(v) = \exp(-A(v)),$$

$$q(v) = \bar{Q}(v) - e(v).$$

Using positive numbers γ, d, L, l, which satisfy the following conditions

$$\begin{aligned} d &< 1 < \gamma, \\ l d^{\gamma} e^{d} &< 1. \end{aligned} \tag{3}$$

Denote $C^1[0, d]$ the set of continuously differentiable functions on the segment $[0, d]$. Define the sets

$$B(\gamma, L, d) = \{q(v) \in C^1[0, d] : |q(v)| \le L v^{\gamma}, v \in [0, d]\},$$

$$\Gamma^+(\gamma, L, l, d) = \{q(v) \in C^1[0, d] : l v^{\gamma} \le q(v) \le L v^{\gamma}, v \in [0, d]\},$$

$$\Gamma^-(\gamma, L, l, d) = \{q(v) \in C^1[0, d] : -L v^{\gamma} \le q(v) \le -l v^{\gamma}, v \in [0, d]\}.$$

Which satisfy the relations

$$\begin{aligned} \Gamma^+(\gamma, L, l, d) &\subset B(\gamma, L, d), \\ \Gamma^-(\gamma, L, l, d) &\subset B(\gamma, L, d). \end{aligned} \tag{4}$$

It is easy to prove the statement which allows to check effectively the belonging of the function q to the classes $\Gamma^+(\gamma, L, l, d), \Gamma^-(\gamma, L, l, d)$.

Lemma B.1 *Assume that for fix $l, L, \gamma, \gamma > 1, l < L$, there are $l_*, l < l_* < L$ (there are $l_*, -L < l_* < -l$) so that the function $q(v)$ satisfies the relation*

$$q(v) \sim l_* v^{\gamma}, v \to 0. \tag{5}$$

Then there is d satisfying the condition (3) so that the function $q(v) \in \Gamma^+(\gamma, L, l, d)$ (the function $q(v) \in \Gamma^-(\gamma, L, l, d)$).

Lemma B.2 *Assume that the function $q \in B(\gamma, L, d)$ then for $n \ge (Le^{a})^{1/(\gamma-1)}$ we have*

$$\Delta_n \le 2 L w_1^{\gamma} n^{1-\gamma}, \tag{6}$$

where w_1 is single solution of the equation

$$L n^{1-\gamma} w^{\gamma} = exp(-aw), \quad w > 0. \tag{7}$$

Proof Assume that $\alpha = 1 - \varepsilon, 0 < \varepsilon < 1$, and $n^{-\alpha} < d$. It is obvious that

$$\begin{aligned}\Delta_n &= \sup_{v\geq 0}|\bar{Q}^n(v) - \exp(-nav)| \\ &= \max\left(\sup_{0\leq v\leq n^{-\alpha}}|\bar{Q}^n(v) - e^n(v)|, \sup_{v\geq n^{-\alpha}}|\bar{Q}^n(v) - e^n(v)|\right) \\ &\leq \max\left(\sup_{0\leq v\leq n^{-\alpha}}|\bar{Q}^n(v) - e^n(v)|, e^n(n^{-\alpha}), \bar{Q}^n(n^{-\alpha})\right).\end{aligned}$$

Put $D_n = \sup(|q(v)| : 0 \leq v \leq n^{-\alpha})$, we obtain

$$\Delta_n \leq \max(nD_n, \exp(-an^{\varepsilon}), \exp(-an^{\varepsilon}) + nD_n),$$

and consequently

$$\Delta_n \leq 2\max(\exp(-an^{\varepsilon}), nD_n). \tag{8}$$

From the condition $q \in B(\gamma, L, d)$ we obtain that $D_n \leq Ln^{-\gamma(1-\varepsilon)} = Ln^{-\gamma}n^{\gamma\varepsilon}$ and so from (8) we have

$$\Delta_n \leq 2\max\left(Ln^{1-\gamma}n^{\gamma\varepsilon}, \exp(-an^{\varepsilon})\right). \tag{9}$$

Fix $n \geq (Le^a)^{1/(\gamma-1)}$, then $w_1 > 1$. Choose $\varepsilon = \varepsilon(n) = w_1^{\gamma}$, from Formula (9) we have Formula (6).

Remark B.2 Estimates from Lemma B.2 earlier were obtained for some other classes of functions in [4]. But these estimates are necessary for further investigation.

Lemma B.3 *Suppose that d satisfies the inequality ln(ln n) < nd,*

1. *If the function* $q \in \Gamma^+(\alpha, L, l, d)$, *then*

$$\Delta_n \geq \frac{l}{\ln n}n^{1-\gamma}. \tag{10}$$

2. *If the function* $q \in \Gamma^-(\gamma, L, l, d)$, *then*

$$\Delta_n \geq \frac{l}{\ln n}n^{1-\gamma}\left(1 - \frac{l}{\ln n}n^{1-\gamma}\right). \tag{11}$$

Proof

1. By definition

$$\begin{aligned}\Delta_n &\geq \sup_{0\leq v\leq d}\left(\bar{Q}^n(v)-e^n(v)\right)\\ &=\sup_{0\leq v\leq d} e^n(v)\left(\left(1+q(v)e^{-1}(v)\right)^n-1\right)\\ &\geq \sup_{0\leq v\leq d} e^n(v)\left(\left(1+le^{-1}(v)v^{\gamma}\right)^n-1\right)\\ &\geq l\sup_{0\leq v\leq d} e^n(v)nv^{\gamma}e^{-1}(v).\end{aligned}$$

Assume that $\alpha=1-\varepsilon, 0<\varepsilon<1, n^{-\alpha}<d$, then

$$\Delta_n\geq l\sup_{0\leq v\leq n^{-\alpha}} e^n(v)nv^{\gamma}\geq \ln^{1-\gamma}\exp(-an^{\varepsilon})n^{\gamma\varepsilon}\geq \ln^{1-\gamma}\exp(-an^{\varepsilon}).$$

Putting

$$\varepsilon=\frac{\ln[\ln(\ln n)]-\ln a}{\ln n} \tag{12}$$

and using the condition *ln(ln n)* < nd, we obtain the inequality (10).

2. Analogously with (1) we obtain

$$\begin{aligned}\Delta_n &\geq \sup_{0\leq v\leq d}\left(e^n(v)-\bar{Q}^n(v)\right)\\ &=\sup_{0\leq v\leq d} e^n(v)\left(1-\left(1-q(v)e^{-1}(v)\right)^n\right)\\ &\geq \sup_{0\leq v\leq d} e^n(v)\left(1-\left(1-lv^{\gamma}e^{-1}(v)\right)^n\right)\\ &\geq \sup_{0\leq v\leq d} e^n(v)nlv^{\gamma}\left(1-lv^{\gamma}e^{-1}(v)\right)^{n-1}.\end{aligned}$$

and so

$$\begin{aligned}\Delta_n &\geq \sup_{0\leq v\leq n^{-\alpha}} e^n(v)nlv^{\gamma}\left(1-lv^{\gamma}e^{-1}(v)\right)^{n-1}\\ &\geq ln^{1-\gamma}\exp(-an^{\varepsilon})n^{\varepsilon\gamma}(1-(n-1)\ln^{-\alpha\gamma}\exp(an^{-\alpha}))\\ &\geq ln^{1-\gamma}\exp(-an^{\varepsilon})n^{\varepsilon\gamma}\left(1-n^{1-\gamma}n^{\varepsilon\gamma}\exp(an^{-\alpha})\right)\\ &\geq ln^{1-\gamma}\exp(-an^{\varepsilon})n^{\varepsilon\gamma}\left(1-n^{1-\gamma}n^{\varepsilon\gamma}\exp(an^{\varepsilon})\right).\end{aligned}$$

From (12) using the condition $ln(ln\,n)<nd$, we obtain the inequality (11).

Theorem B.3

1. *If the conditions* (1) *of Lemma B.3 are true then two sided estimate*

$$\frac{l}{\ln n} n^{1-\gamma} \leq \Delta_n \leq 2Lw_1^{\gamma} n^{1-\gamma}.$$

   takes place.

2. *If the conditions* (2) *of Lemma B.3 are true then two sided estimate*

$$\frac{l}{\ln n} n^{1-\gamma}\left(1 - \frac{l}{\ln n} n^{1-\gamma}\right) \leq \Delta_n \leq 2Lw_1^{\gamma} n^{1-\gamma}.$$

   takes place.

Proof The statements of Theorem 3 are corollaries of Lemmas B.1, B.2 and B.3.

B.4. Conclusion

Two sided estimates of distances between prelimit and limit distributions in the scheme of r.v.'s minimum with identical power multipliers by n and with different logarithmic multipliers are obtained. These results are adequate as with Weibull so with Gompertz limit distributions for life time distributions in stochastic systems.

More detailed considerations of limit distributions in schemes of random vectors minimum and maximum including Marshall-Olkin d.f.'s are in [5].

References

[1] Rocchi, P. (2002). Boltzman like entropy in reliability theory. *Entropy, 4*, 142–150.

[2] Rocchi, P., & Tsitsiashvili, G. Sh. (2004). About reversibility and irreversibility of stochastic systems. *Proceedings of International Conference on Foundations of Probability and Physics*-3, *AIP Conference Proceedings Series, 750*, 340–350.

[3] Gnedenko, B. V. (1988). *Course of probability theory*. Moscow: Science (In Russian).

[4] Siganov, I. S. (1983). Several remarks on applications of one approach to studies of characterization problems of Polya theorem type. *Proceedings of the 6-th International Seminar. Lecture Notes in Mathematics*, 227–237.

[5] Tsitsiashvili, G. Sh., & Rocchi, P. (2014). Limit theorems and stochastic entropy properties of Marshall-Olkin distributions. *Journal of Mathematical Sciences, 196*(1), 97–101.

Appendix C
Mortality Plateau

A phenomenon, which is typical of biological systems and seems to be alien to engineering systems, is the deviation of the mortality rate from the ideal law at advanced age. Various scholars, including Gompertz, were aware of how the exponential-exponential distribution does not continue past age 80 in humans (see Fig. 6.11). In 1939 Greenwood and Irwin first discussed the *late-life mortality deceleration* (also called *mortality leveling-off* and even the *late-life mortality plateau*) [1]. This still represents a challenging question for researchers in various fields, in that the accurate knowledge of human mortality is important for economics, taxation, insurance, medicine etc.

This appendix means to offer a contribution to the discussion; it follows the deductive logic and uses the results obtained in the previous chapters. We call this proposal the "*theory of the simplified structure*" due to the theorem proved below.

C.1. Experimental Evidence

It is well known that the structure of biological organs reduces in complexity at advanced age [2]. The following list exhibits some of the most evident effects which Makeham defined as "diminution of the vital power" [3]:

1. Lean body mass diminishes,
2. Muscles' strength reduces,
3. The number of nerve fibers decreases,
4. Bones are frail because of reduced bone mineral density,
5. The number of living cells in the cerebral cortex of the brain decrease,
6. Sense organs and other internal organs show evident hypertrophy,
7. Energy reserves are in short supply,
8. Skin becomes thin,
9. Reaction to the stimuli delays,
10. Some functions completely vanish such as the reproductive function.

P. Rocchi, *Reliability is a New Science*, DOI 10.1007/978-3-319-57472-1

We separate the *normal aging period* (NAP)—complying with the Gompertz-Makeham law—from the *advanced aging period* (AAP) when the phenomena 1–10 and even other effects, begin to be seen and reduce the form of *S*.

C.2. Simplification Assumption

The signs of aging are global, they demonstrate how the entire structure of *S* becomes less complicated at advanced age. The *physical model* of organisms simplifies when they pass from NAP to AAP; it changes from the regular intricate shape toward a more linear one.

C.2.1 The complex cascade effect (discussed in Chap. 6) progressively reduces in complexity in consequence of the general simplification described above. It may even fade away due to the feebler interactions amongst functions which weaken. On the basis of the previous chapters, we should take into account three structures that occur during AAP in the order as follows:

- S_A encompasses the compound cascade effect,
- S_B encompasses the linear cascade effect, (1)
- S_C does not encompass any cascade effect.

The reduction of the human body does not occur all of a sudden, but in progressive manner, and thus the sub-systems S_A, S_B and S_C cooperate in a way during AAP and assumption (1) may be formally translated as

$$S = (S_A \; AND \; S_B \; AND \; S_C). \tag{2}$$

Theorem C.1 or Theorem of the Simplified Structure *Let t_i the initial time of the advanced aging period, the mortality rate $\lambda(t)$ is the following*

$$\lambda(t) = a \exp(t) Q(t) + bt^n R(t) + cT(t), \quad t \geq t_i; a, b, c > 0. \tag{3}$$

Where Q(t), R(t) and T(t) are weight-functions that vary in (0,1) and obey the constraint

$$[Q(t) + R(t) + T(t)] = 1, \qquad t \geq t_i. \tag{4}$$

Proof From (2.5) we have

$$P_f = (P_{fA} \cdot P_{fB} \cdot P_{fC}). \tag{5}$$

The previous chapters imply that P_{fA} is to be calculated using Theorem 6.5, P_{fB} by means of Theorem 6.3 and P_{fC} by Theorem 4.1. For the sake of simplicity, suppose that the constants a and b in (6.8) equal the unit, g and d in (6.18) be equal to the unit, thus one gets

$$\begin{aligned} P_f(t) &= P_{fA} \cdot P_{fB} \cdot P_{fC} \\ &= \exp[-a \exp(t)] \cdot \exp(-bt^n) \cdot \exp(-ct) \\ &= \exp\ -\ [a \exp(t) + bt^n + ct], \qquad a, b, c > 0. \end{aligned} \tag{6}$$

From the definition of hazard function (1.5) we obtain

$$\lambda(t) = -\frac{P'(t)}{P(t)} = a \exp(t) + nb\,t^{n-1} + c. \tag{7}$$

For the sake of simplicity we rewrite (7) this way

$$\lambda(t) = a \exp(t) + b\,t^n + c, \quad a, b, c > 0. \tag{8}$$

This demonstration is incomplete since the subsystem S_B (and also S_C) does not exist in NAP and begins to expand at the expense of S_A in AAP. The larger becomes S_B and the smaller is S_A by time passing. This means that the contribution of the exponential term in (8) diminishes with time while the contribution of the power term increases and the linear term is null. S_B in turn decrease; the cascade effect ceases to exist and the sub-system S_C begins to substitute S_B. The accurate description of the mortality rate should include the *weight-functions* $Q(t)$, $R(t)$ and $T(t)$ which respectively specify the contribution of each sub-systems S_A, S_B and S_C by time passing. Each weight-function varies between zero and the unit

$$0 \le Q(t), R(t), T(t) \le 1.$$

A weight-function gives an addend more "weight" or influence on the result than other addends of the sum. Thus the mortality rate (8) has this form

$$\lambda(t) = a \exp(t)Q(t) + bt^n R(t) + cT(t), \quad t \ge t_i; a, b, c > 0.$$

Each weight-function is to be lower than the unit because of (5) while the summation equals the unit

$$[Q(t) + R(t) + T(t)] = 1, \quad t \ge t_i.$$

C.2.2 As soon as the organs progressively lose vigor, S swaps from S_A to S_B, the power term [$=bt^n$] prevails over the exponential term [$=a\exp(t)$] and the curve of the mortality force flexes. Later on any cascade effect vanishes, S_B ceases to exist and S becomes S_C, $\lambda(t)$ coincides with the linear term and the curve becomes flat.

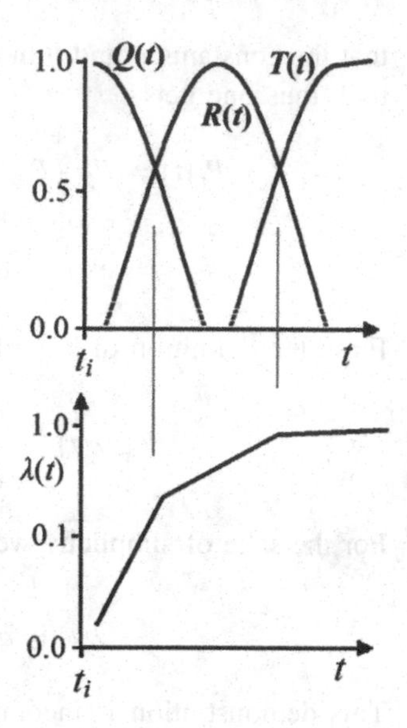

Fig. C.1 Visual relationships between the weight-functions (*top*) and the resulting mortality deceleration (*bottom*)

By way of illustration, suppose $Q(t)$, $R(t)$ and $T(t)$ are three curves as in Fig. C.1 (*top*). The mortality rate curve progressively flattens because of the effect of the addends which gain weight in turn.

Example: Thatcher, Kannisto and Vaupel [4] have studied the demographic data of thirteen countries which have a sufficiently long series of data to be useful for analyzing the range of ages from 80 to 120. There is data for the cohorts born in 1871–80 (A), for the period 1960–70 (B), the period 1970–80 (C) and the period 1980–90 (D). The populations are subdivided into male and female. Figure C.2 plots the mortality force of all the women in the thirteen countries which is in good agreement with the present theory.

C.3. Comments on Theorem C.1

The late-life mortality deceleration is a controversial issue because of the obstacles encountered at the experimental and theoretical levels.

C.3.1 Statistical studies of extreme longevity are difficult since few individuals live to very old ages. Secondly, historical series are lacking. Thirdly, empirical data collected in databases covering several countries—e.g., the International Database on Longevity (IDL), the Berkley Mortality Database and the Kannisto and Thatcher Database (K-T Database)—does not provide uniform accounts of the human mortality plateau [5] Robine. Statisticians find distributions whose tails are in opposite directions [4] Thatcher et al. Figure C.3 exhibit two example cases which diverge from the present theory.

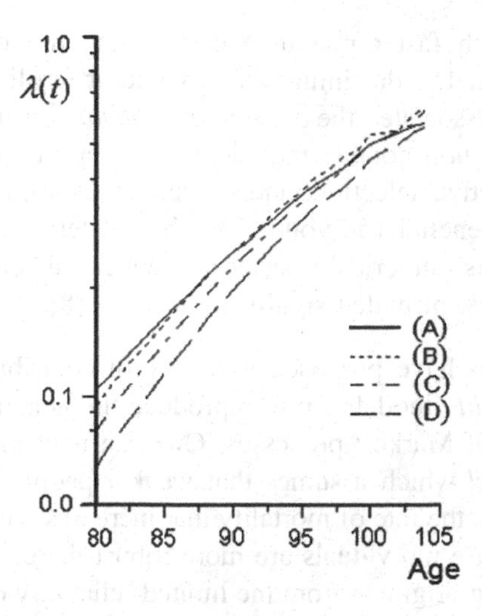

Fig. C.2 The force of mortality of all the women Reproduced from [4] with permission from University Press of Southern Denmark

There is no doubt that the mortality rate flexes after one hundred years of age but the data of some cohorts show differing pictures. The mortality of animals at advanced age appears even more varied and multifold.

C.3.2 Biologists put forward an assortment of theories to explain late-life mortality deceleration. Evolutionary genetic studies offered the first key to explanation. Peter Medawar [6] supposed that over evolutionary time late-acting mutations will

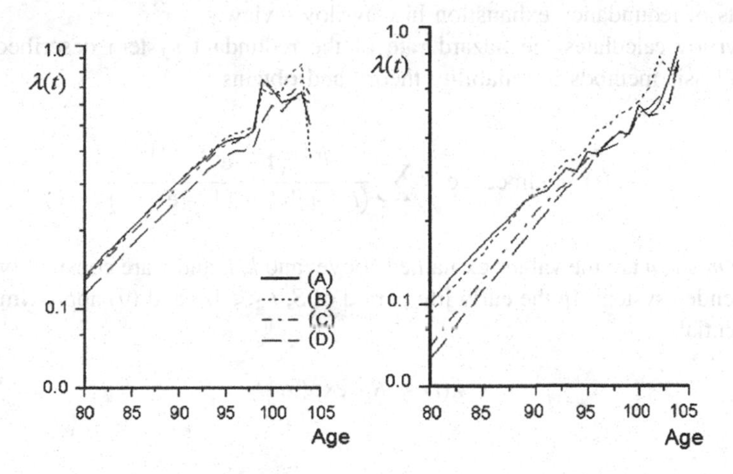

Fig. C.3 The force of mortality of West Germany male (*left*) and Denmark female (*right*) Reproduced from [4] with permission from University Press of Southern Denmark

accumulate at a much faster rate than early-acting mutation. These late-acting mutations will thus lead to declining viability and/or fertility as an organism ages. George C. Williams associated the *antagonistic pleiotropy* hypothesis to the theory of *mutation accumulation* notably mutations affecting old ages accumulate thanks to a weaker reproductive selection, and the antagonistic pleiotropy reinforces this effect since a gene, beneficial in youth, may be deleterious in old age [7].

Computer scientists numerically tested the genetic theories. The *Penna model*, devised in the nineties, provided significant support [8].

C.3.3 Mathematicians have proposed a variety of contributions to the reliability theory. *Changing-frailty* models try to reproduce the general features of mortality curves on the basis of Markov processes. One can read the *Vaupel, Manton and Stallard (VMS) model* which assumes that weaker people tend to die earlier [9]. From this assumption, the rate of mortality that increases with age should decline at advanced ages, because individuals are more robust there. Koltover holds that the general laws of ageing originate from the limited reliability of biological systems at all functional levels, starting from the level of enzymes [10]. An ample study was conducted by Gavrilov and Gavrilova [11]. They assume that organisms form themselves in ontogenesis through a process of self-assembly out of untested elements. These authors emphasize the extraordinary degree of miniaturization of the biosystem components based on cellular units, and they conclude that aging results in the reduction of redundant elements. Gavrilov defines the suitable *physical model* in this way. The system *S* includes *m* serially connected blocks, each being critical for system survival. Each block has *n* mutually substitutable elements even called 'redundant elements'. Intrinsic and extrinsic phenomena randomly knock out the redundant elements and the number *n* drops down in each block with time. As defects accumulate, the redundancy in the number of elements eventually disappears and the organism degenerates into a structure with no redundancy. Aging consists of redundancy exhaustion in Gavrilov's view.

Gavrilov calculates the hazard rate of the redundant system described above using classic methods in reliability theory and obtains

$$\lambda(t) = \mathrm{klmce}^{-\lambda}\mathrm{e}^{-kt}\sum_{x=1}^{n}\frac{l^{x-1}\left(1-\mathrm{e}^{-kt}\right)^{x-1}}{(i-1)!(1-(1-\mathrm{e}^{-kt})^{x})}. \tag{9}$$

Where *m* and *n* are the values explained above; and *k, l,* and *c* are constant typical of the intended system. In the early life period of *S*, $t << 1/k$ and (9) approximates the exponential

$$\lambda(t) \approx R \cdot \exp(\alpha t).$$

Where R is a constant. In the late lifetime period $t >> 1/k$ and (9) approximates a constant

$$\lambda(t) \approx mk.$$

Mortality decelerates with subsequent leveling-off, as an inevitable consequence of redundancy exhaustion.

Factually each degenerative phenomenon from 1 to 10—placed in the initial list—entails the decreasing redundancy of cells and tissues. The present work focuses on the linearization of the system, while Gavrilov focuses on the reduced redundancy of organs. The two mechanisms do not contradict each other and Fig. C.4 shows how a system that is non-redundant from the Gavrilov perspective becomes linear. It is intriguing that Gavrilov's scheme and the present book adopt similar *physical models*, they develop different methods of calculus and reach consistent conclusions.

It is also necessary to recall a further achievement of Gavrilov and Gavrilova who demonstrate that the relative differences in the mortality rates of compared populations (within a given species) decrease with age, and mortality convergence is observed due to the exhaustion of initial differences in redundancy levels. They call this interesting result '*the compensation law of mortality*'.

C.3.4 The *principle of superposition*—the first law of the present theory—entails that the system end depends on many causes.

There is a large number of mechanisms that reduce the 'vital power' of a living being at advanced age. The effects listed from 1 to 10 offer a few examples. In addition there are the antagonistic pleiotropy, the mutation accumulation, the more robust people theory (VMS theory) and other mechanisms which run in parallel. The variety of determinants which lead to the mortality deceleration do not cancel one another out from the perspective of the *superposition principle*. The multiple dependent competing failure processes do not start all together no do they have identical impact. The 'orchestra' encompassed by these processes can but prompt differing trends in various cohort data and is able to explain the assortment of

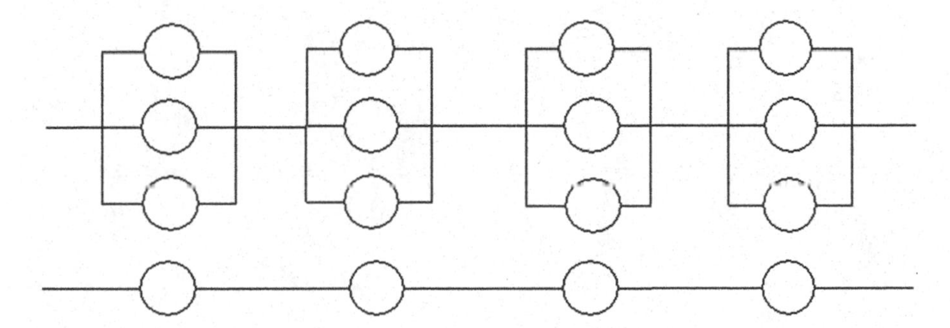

Fig. C.4 Redundant system with $m = 3$ and $n = 4$ (*top*); Non-redundant system with $m = 1$ and $n = 4$ (*bottom*)

distributions found at advanced age by statisticians. This inclusive and complete view provides a justification for the different trends of mortality during advanced age.

References

[1] Greenwood, M., & Irwin, J. O. (1939). The biostatistics of senility. *Human Biology, 11*, 1–23.
[2] Harman, D. (1981). The aging process. *Proceedings of the National Academy of Sciences, 78* (11), 7124–7128.
[3] Timiras, P. S. (ed.) (2007). *Physiological basis of aging and geriatrics*. CRC Press.
[4] Thatcher, A. R., Kannisto, V., & Vaupel, J. W. (1999). *The force of mortality at ages 80–120*. Monographs on Population Aging, 5, Odense University Press.
[5] Robine, J. M. (2006). Research issues on human longevity. *Human Longevity, Individual Life Duration, and the Growth of the Oldest-Old Population, 4*, 7–42.
[6] Medawar, P. B. (1952). *An unsolved problem of biology*. H. K. Lewis Publ.
[7] Mueller, D. L., & Rose, M. R. (1996). Evolutionary theory predicts late-life mortality plateaus. *Proceedings of the National Academy of Sciences USA, 93*, 15249–15253.
[8] Coe, J. B., Mao, Y., & Cates, M. E. (2002). A solvable senescence model showing a mortality plateau. *Physical review letters, 89*, 288103.
[9] Vaupel, J. W., Manton, K. G., & Stallard, E. (1979). The impact of heterogeneity in individual frailty on the dynamics of mortality. *Demography, 16*, 439–454.
[10] Koltover, V. K. (1997). Reliability concept as a trend in biophysics of aging. *Journal of Theoretical Biology, 184*, 157–163.
[11] Gavrilov, L. A., & Gavrilova, N. S. (2001). The reliability theory of aging and longevity. *Journal of Theoretical Biology, 213*, 527–545